Generis
PUBLISHING

Beyond Compliance: Designing Safety Standards for Human-Centered Innovation

Lisa Giusti Gestri

Title: **Beyond Compliance:
Designing Safety Standards for
Human-Centered Innovation**

ISBN: 979-8-89966-278-2

Author: Lisa Giusti Gestri

Cover image: Dino Gestri

Publisher: Generis Publishing
Online orders: www.generis-publishing.com
Contact email: info@generis-publishing.com

Beyond Compliance: Designing Safety Standards for Human-Centered Innovation

Back Cover Bio

Dr. Lisa Giusti Gestri is a designer, researcher, and advocate for human-centered innovation. With experience across Italy and Australia, she explores how safety, usability, and design intersect to protect lives and unlock innovation. From jockeys' vests to industrial products, her work challenges outdated standards and imagines smarter, more ethical systems built around people, not rules.

How can safety standards keep pace with human needs and design innovation?

This book reveals how comfort, usability, and inclusivity can be embedded into safety standards to improve performance and protect lives. Through the lens of jockeys' safety vests in Australian horse racing, it introduces a new framework for understanding user interdependence and shows how outdated regulations can block design innovation and progress. Blending design research, technology, and policy, it offers a bold vision for safety systems that are smarter, more ethical, and truly human-centered.

Acknowledgments

This book would not have been possible without the generous support, guidance, and insights of many people. A special thanks to the editors for their feedback and advice.

I am deeply grateful to the jockeys who shared their experiences, the medical professionals who provided critical perspectives, and the industry stakeholders who welcomed me into their world. Your voices shaped the heart of this work.

Also, I would like to thank my academic mentors and collaborators for their invaluable guidance throughout my doctoral research and beyond. Your expertise, encouragement, and rigorous standards inspired me to pursue this project with care and precision.

Last but not least, to my parents, my brother, and my grandparents, whose unwavering support, wisdom, and encouragement have shaped the person and scholar I am today. I am grateful for your quiet strength and boundless love, which have carried me through every challenge. Thank you for planting the seeds of curiosity, courage, and resilience that bloom in this work. To my husband, thank you for your patience, belief, and love, which sustain me every day. And to my son, your curiosity, joy, and presence remind me of the wonder and purpose in all that I do. Your laughter, bravery, and fearless heart illuminate me every day: you are the living reminder of why this journey matters. This book is as much yours as it is mine, born from the love and inspiration you all have given me.

About the Author

Author: Lisa Giusti Gestri
ORCID: https://orcid.org/0000-0001-6640-1067

Dr. Lisa Giusti Gestri is an industrial and product designer, design researcher, and academic with extensive experience across Italy and Australia. Her research bridges industrial design, product standards, and safety innovation, with a focus on high-impact sports and human protection. She is passionate about leveraging design to improve people's lives. Creating and supervising projects from ideation to completion on-site has given her vast knowledge in research, design thinking, human-centered design, innovation, advanced materials, industrialized buildings, and user experience design. Her diverse experience makes Lisa a versatile professional at the intersection of design and technology. She explores how human-centered design can transform outdated regulations and improve lives. She holds a PhD and MR-DES from Swinburne University of Technology and a BIndDes from the University of Florence.

Synopsis

How can safety standards keep pace with human needs and design innovation?

This book shows how core design values (e.g., comfort, usability, inclusivity, and responsiveness) can be embedded into safety standards to improve performance, innovation, and user experience. Using the Australian horse racing industry as a case study, it introduces the Dependency-Based User Experience (D-UX) framework, revealing how both primary users, like jockeys, and co-dependent users, such as medical staff and safety officers, shape the effectiveness of protective design. The D-UX framework is versatile and can be adapted across industries where multiple users rely on the same product design, ensuring safety and usability for all.

Bridging design research, technology, and policy, the book examines how existing safety standards can enable or constrain innovation. It proposes conceptual redesigns and offers evidence-based recommendations for reforming standard-making processes to reflect real-world user interdependencies. Grounded in original doctoral research and enriched by contemporary scholarship, this book provides a roadmap for creating safety systems that are robust, ethical, and truly human-centered.

Preface

This book is the culmination of a journey that began with a simple yet pressing question: Why do safety vests for jockeys (designed to protect lives) often feel so poorly suited to the people who wear them?

As a design researcher and industrial designer, I have spent years exploring the intersection of product standards, user experience, and innovation in high-risk sports. My doctoral research, conducted with the support received from Racing Victoria, Racing Australia, and the Australian Jockeys' Association with the National Jockeys' Trust, , uncovered a profound disconnect between the technical criteria embedded in safety standards and the lived realities of those who rely on protective gear.

Through interviews, fieldwork, and design prototyping, I came to understand that safety is not just about compliance: it is about care, trust, and dignity. This book builds on that insight, introducing a new framework for designing product standards that centers on core design values and co-dependent user experience.

It is written for designers, regulators, manufacturers, and anyone who believes that personal protective equipment (across any field) should not only protect but also empower.

I am deeply grateful to the jockeys, medical professionals, and industry stakeholders who generously shared their experiences and insights. Their voices shaped this work. This book is a call to critically rethink how safety is conceptualized, designed, and defined.

Dr Lisa Giusti Gestri

Table of Contents

List of Tables

List of Figures

List of Acronyms and Abbreviations

ARB:	Australian Racing Board
ASTM:	Advancing Standards Transforming Markets (formerly known as American Society for Testing and Materials)
BETA:	British Equestrian Trade Association
BSI:	British Standards Institution
CEN:	European Committee for Standardization
D-UX:	Dependency-based User Experience
EN:	European Norm
EU:	European Union
HCD:	Human-Centered Design
ISO:	International Organization for Standardization
JRA:	Japan Racing Association
OECD:	Organization for Economic Co-Operation and Development
OPSI:	Observatory of Public Sector Innovation
PPE:	Personal Protective Equipment
PSI:	Public Sector Innovation
SATRA:	Shoe and Allied Trades Research Association
SHESC:	Swinburne Human Research Ethics Committee
UCD:	User Centered Design
UK:	United Kingdom
UXD:	User Experience Design
WTO:	World Trade Organization

Chapter 1.
Introduction – Why Design Values Matter in Product Standards

Design is rarely neutral. Every product embodies choices (about feeling, needs, safety, usability, performance, pleasure of using it, and aesthetics) that reflect underlying values. In regulated environments, such as high-risk sports or industrial settings, those choices are constrained by standards and compliance requirements. Yet, standards themselves are not value-free; they carry assumptions about risk, performance, and the priorities of those who develop them.

In the context of thoroughbred horse racing, the stakes are literal. Jockeys are highly trained professional athletes who ride racehorses at high speeds, often exceeding 60 km/h, in competitive and unpredictable conditions. Their role demands agility, precision, and endurance, while exposing them to significant risks, including falls, collisions, and serious injuries. Personal protective equipment (PPE), such as safety vests, is intended to mitigate these risks, yet poorly designed gear can hinder performance, cause discomfort, or even increase danger in critical situations. Consequently, jockeys are mandated to wear protective equipment that reduces the risk of injury while supporting, rather than hindering, athletic performance. Figure 1 illustrates an example of the approved gear. Although regulatory specifications differ depending on the rider's age and the type of event, the overarching purpose of this PPE is to safeguard the jockey while maintaining the functional demands of competitive racing.

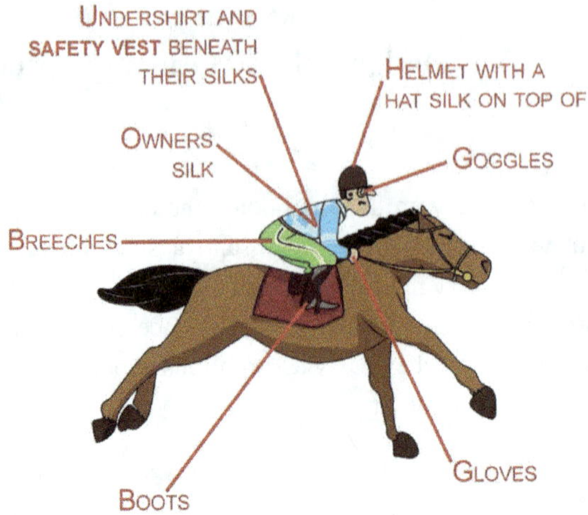

Figure 1. Standard jockeys' PPE equipment required for races (author's work)

Before the thunder of hooves and the cheer of the grandstand, there's a hush, a quiet moment shared between a jockey and their horse. In that pause, you can almost feel the energy of what's to come. Jockeys are remarkable athletes, though you might not guess it from their slight frames. Every morning, before most people have even stirred from bed, they're already at the stables, riding through the cool dawn air. The mist rises off the track, and the only sounds are hoofbeats and the steady rhythm of breathing, horse and rider in perfect sync. A jockey's life is a demanding one. It takes strength to control half a ton of muscle at full speed, and discipline to maintain the strict weight limits that come with the profession. There's little glamour behind the scenes: early mornings, long hours, and the constant challenge of staying fit, focused, and in tune with every horse they ride. Yet for most jockeys, the reward is not in the trophies, but in the connection, that fleeting, wordless bond between human and horse that makes every ride feel alive.

Across Australia, thoroughbred racing isn't just a sport; it's part of the national identity. From the polished spectacle of the Melbourne Cup, where millions pause to watch, to the small-town country meets that bring local communities together, racing has a rhythm all its own. The industry stretches from sprawling breeding farms in the Hunter Valley to remote outback tracks lined with red dust and gum trees. It supports thousands of people (e.g., breeders, trainers, stable helpers, farriers), and, of course, the jockeys who bring the horses' power to life. In recent years, the Australian racing community has placed growing emphasis on welfare and sustainability, ensuring the well-being of both horses and riders. It's a reminder that the sport's true spirit lies not just in competition, but in respect for the animals, for the craft, and for the generations who have kept this tradition alive. In the end, every race tells a story of patience, courage, and partnership. And at the center of it all is the jockey: a figure of balance, grit, and quiet determination, steering both horse and history down the long, bright stretch of the track.

This chapter explores the critical role of design values in shaping product standards, emphasizing that thoughtful design can reconcile regulatory compliance with human-centered innovation. Through the lens of safety vests for jockeys, we examine how standards can simultaneously enable and restrict design, influencing both the experience of users and the possibilities for innovation. By understanding this interplay, designers, regulators, and manufacturers can create products that are not only safe but also functional, dignified, and empowering for those who rely on them. Hence, it begins with a detailed case study of Australian jockey safety vests, incorporating insights from both the athletes and medical professionals involved. This example highlights how design values manifest in practice and illustrates the potential for human-centered approaches to improve compliance, usability, and overall safety.

1.1 Rethinking Safety Through Design

In a world increasingly shaped by technology, regulations, and user expectations, the role of design extends far beyond aesthetics. Design values, the principles that guide how products are conceived, developed, and experienced, play a crucial role in shaping not only the form and function of products but also the standards that govern them. Standards are often seen as rigid rules, yet they also reflect the priorities, assumptions, and values of the communities and industries that create them. By examining the intersection of design values and product standards, this chapter highlights how thoughtful design can bridge the gap between safety, performance, and user experience. Through a human-centered lens, designers can ensure that compliance does not mean compromise and that innovation remains a driving force even within regulated environments. The following sections explore these ideas in practice, beginning with the first case study of Australian jockey safety vests (involving both jockeys and related medical professionals), to demonstrate how standards can both challenge and inspire meaningful design solutions, but as per this case, may also be restrictive in introducing innovation (Giusti Gestri, 2019). Understanding how design values shape the development of product standards is crucial for creating solutions that are both innovative and user-centered. The author's case study highlights the complex relationship between regulatory compliance and design innovation.

In the twenty-first century, products are shaped as much by regulatory requirements and social expectations as by technological advancement. Within this context, design cannot be understood solely as a matter of form or aesthetics. Instead, it represents a set of values and priorities that influence how products are conceived, developed, and experienced. Design values, principles such as safety, usability, empathy, and inclusivity, are not abstract ideals. They determine how products function in practice and, importantly, how they are codified within product standards.

Product standards are typically regarded as rigid, technical frameworks: codified requirements that guarantee safety, reliability, and interoperability. Yet standards are not neutral. They are cultural artifacts, embodying the assumptions and priorities of industries, regulators, and technical experts. When a standard specifies minimum impact resistance for a protective vest or defines the acceptable range of material properties, it reflects collective decisions about what matters most: durability, affordability, safety margins, or manufacturing efficiency. Advances in information systems and emerging technologies are reshaping contemporary society. As a result, embedding ethical and moral considerations into the conception, design, and implementation of these technologies has become increasingly important (Manders-Huits, 2011). It is equally vital to critically examine the underlying value orientations and assumptions that guide specific design choices. Innovation serves as a central driver for improving both product performance and user benefits, positioning designers as key actors in envisioning and implementing solutions that respond to evolving contexts of use and shifting customer demands.

In the aftermath of the COVID-19 pandemic, lifestyles and behavioral patterns have undergone substantial transformation. In particular, the healthcare sector has been compelled to extend services into domestic environments, effectively turning homes into sites of care delivery (Perego & Sironi, 2022). Consequently. technology has assumed a more critical role in addressing these evolving needs, with the goal of enhancing not only quality of life but also the safety and well-being of both patients and caregivers. Hence, a Human-Centered Design (HCD) approach is essential: one that considers not only performance and safety but also emotional experience and user needs. While standards aim to ensure consistency and safety, they can sometimes limit creativity, novelty, and material options. To address this tension, strategies like co-design provide a way forward, allowing designers to include user insights and develop standards that support, rather than hinder, meaningful innovation. Design standards play a critical role in shaping the products we interact with daily, especially in high-risk industries where safety is paramount. Yet, the process of developing these standards often prioritizes technical specifications and regulatory compliance over nuanced user experiences. This disconnect can result in products that meet formal requirements but fall short in addressing real-world needs, preferences, and emotional responses. In the context of sports' safety equipment (such as jockey vests), this gap becomes particularly evident, where the stakes involve not just comfort or usability, but users' lives and safety.

Product standards play a pivotal role in shaping the design, development, and lifecycle of products across industries. They serve as codified benchmarks that ensure safety, reliability, and interoperability, thereby fostering consumer trust and regulatory compliance (Riganelli, 2025). In engineering contexts, standards provide designers with structured methodologies, material specifications, and performance criteria, which streamline the design process and reduce ambiguity (IEEE, 2018). The integration of standards into product design enhances development efficiency by offering proven frameworks and reducing the need for iterative trial-and-error approaches. This not only accelerates time-to-market but also minimizes costly redesigns and compliance failures (Riganelli, 2025). Moreover, standards facilitate modularity, component commonality, and platform-based design strategies, which are essential for scalable manufacturing and product variety management (Perera, 2006). From a lifecycle perspective, design standards influence downstream processes such as manufacturing, documentation, and data exchange. For instance, standardized documentation protocols ensure that product information is accurately communicated across departments and stakeholders, enabling seamless transitions between design, production, and service phases (Schilli & Kern, 2006). Specifically, in collaborative engineering environments, data exchange standards like ISO STEP (ISO 10303) are critical for interoperability between systems and organizations (Schilli and Kern, 2006). Furthermore, the specification and management of product requirements, often governed by industry standards, are foundational to successful product development. These requirements guide the design intent and ensure alignment with user needs, legal mandates, and market expectations (Göhlich, Bender, Fay, and Gericke, 2021). As such, standards are not merely technical constraints but strategic tools that enhance product quality, reduce risk, and support innovation. Usually, the challenge lies in reconciling the rigidity of standards with the fluidity of design thinking. Standards are typically developed through consensus-driven processes involving regulators, manufacturers, and technical experts. While this ensures consistency and accountability, it can also marginalize the voices of end users, those who ultimately rely on these products for protection. By integrating design values into the standardization process, we can begin to shift the focus from mere compliance to meaningful impact.

1.1.1 The case of Australian jockeys' safety vests

This book explores how design values (such as empathy, usability, emotional engagement, and contextual relevance) can be systematically assessed and embedded into product standards. Drawing on case studies, including the development of safety vests for jockeys in Australia, it highlights the importance of HCD and User Experience Design (UXD) in shaping standards that are not only technically sound but also socially and ethically responsible. Through this lens, the book advocates for a more inclusive and iterative approach to standardization, one that embraces co-design, participatory research, and interdisciplinary collaboration. By doing so, it aims to contribute to a growing body of knowledge that reimagines standards as dynamic tools for innovation, rather than static barriers to progress.

Safety-centered design in high-risk sectors such as sports introduces unique challenges and opportunities, especially concerning the use of (sports) PPE. Despite the general trend toward incremental design innovation, noteworthy benefits can arise from radical approaches, particularly when responding to evolving user demands and regulatory constraints. In high-risk industries such as professional thoroughbred horse racing, safety is often framed through technical compliance and regulatory enforcement. Yet, the design of safety equipment, particularly personal protective gear such as jockeys' safety vests, remains a largely underexplored domain in terms of its experiential, ergonomic, and cultural dimensions. This book argues that design values, the principles, priorities, and user-centered insights that shape how products are conceived and experienced, must be central to the development and revision of every category of product standards. Hence, the case of Australian jockeys' safety vests offers a compelling lens through which to examine this issue. Despite the mandatory use of safety vests since 1998, injury rates among jockeys remain high, and the design of these vests has seen minimal innovation. The Australian standards that govern them (e.g., EN 13158 and ARB 1.1998) have not evolved to accommodate users' needs, advances in materials, wearable technologies, or the changing demographics of users, including the rise of female jockeys.

Design thinking has gained recognition across many fields, yet its application in PPE for athletes (particularly jockeys) remains limited. This chapter emphasizes the importance of interpreting standards through the lens of design values. While regulatory compliance is essential, it alone does not ensure that products effectively meet the needs of users. As illustrated by the first case study on Australian jockey safety vests (Giusti Gestri, 2019), adherence to existing standards has not eliminated injuries nor stimulated significant innovation in vest design. Despite being mandatory since 1998, these vests have undergone minimal design evolution, with limited attention to ergonomics, comfort, or the increasing presence of female jockeys. Standards such as EN 13158 and ARB 1.1998 have struggled to reflect advances in materials, wearable technology, or the broader demographic realities of the sport.

The jockey vest example underscores a wider issue: when standards are treated as static technical checklists, they risk becoming barriers to innovation rather than catalysts for progress. By embedding design values into the standardization process, safety and innovation can be harmonized rather than viewed as competing priorities. Human-centered design approaches, which emphasize lived experience alongside technical performance, offer a pathway to align these goals. Drawing on qualitative and user-experience research, the author developed a dependency-based user experience framework to guide design improvements. The chapter concludes by calling for further research and innovation to overcome the constraints imposed by standards, enabling meaningful advances in design.

1.2 From Compliance to Experience

Traditional personal PPE standards focus on measurable, quantifiable criteria: coverage area, material durability, and impact resistance. These benchmarks are crucial, but they represent only one dimension of safety. The effectiveness of protective gear also depends on how it is experienced by its users. Notably, to preserve athletes' safety during their activities, sports PPE is still considered the best answer. High-impact sports are often defined by their elevated injury risks, making protective equipment a central concern for athletes and regulators alike. Over the past two decades, many disciplines have made substantial progress: cycling has revolutionized helmet technologies, rugby has introduced advanced headgear and padding, and motorsport has adopted body armor and head-support systems that set new global benchmarks for safety. These advancements have been driven not only by stricter standards but also by interdisciplinary collaboration and continuous investment in design innovation.

In stark contrast, equestrian sports (particularly professional horse racing) continue to record some of the highest injury rates in competitive athletics, yet their protective gear has evolved at a much slower pace. Jockey safety vests, mandatory since the late 1990s, remain largely unchanged despite advances in materials science, wearable technologies, and ergonomics. The regulatory standards that govern them, while effective in ensuring minimum levels of compliance, have inadvertently created a barrier to more radical forms of innovation. This situation presents a critical problem: when product standards are treated primarily as technical checklists rather than dynamic frameworks shaped by design values and user experience, they risk limiting the very innovation they are meant to inspire.

The case of jockey safety vests therefore provides a powerful lens through which to examine how design, regulation, and user needs intersect, and, in many cases, collide in shaping the future of safety-critical equipment. Professional jockeys compete at extreme speeds while operating at the limits of their physiological capacity in one of the most high-risk sports worldwide. On top of this, they must adhere to strict weight requirements (typically between 50 and 60 kg in flat racing) placing additional strain on their health and performance. The occupational hazards are considerable: jockeys are routinely exposed to high-velocity impacts, falls, and cumulative strain injuries. Research underscores the severity of these risks. An American study reported injury rates of 600 per 1,000 jockey-years, while an Irish investigation (O'Connor et al., 2017) revealed that flat jockeys sustain, on average, a fall every 250 rides, with more than a third of these incidents (35%) resulting in injury. Consequently, professional jockeys repeatedly report discomfort, restricted mobility, and poor fit when using standardized vests. For female jockeys, in particular, standard sizing systems—historically designed for male body morphologies, create additional challenges (Loibner et al., 2019). Such shortcomings are not trivial inconveniences. Poor ergonomics can discourage proper use, increase fatigue, or even contribute to accidents. In other words, compliance with technical standards does not guarantee real-world safety.

The perspective of secondary users further exposes the limitations of compliance-driven design. Medical professionals, for instance, often struggle to remove jockey vests during emergency treatment. While vest structures are optimized for resistance and durability, they rarely account for rapid accessibility under pressure, leading to dangerous delays in trauma care (Horberry & Burgess-Limerick, 2015).

Traditional product standards prioritize measurable performance criteria: e.g., impact resistance, coverage area, and material durability. While these are essential, they often exclude the lived experience of users .Hence, this book advocates for a paradigm shift from compliance-driven design to value-driven design, where standards are not solely dictated by technical specifications but are informed by user needs, co-dependent user interactions, and contextual factors. To operationalize this shift, it introduces the concept of Dependency-Based User Experience (D-UX) (Giusti Gestri, 2019). D-UX expands the scope of design evaluation to include secondary and tertiary users (such as paramedics, trainers, and safety officers) whose interactions with the product are essential to its function and lifecycle but are often overlooked in traditional design frameworks.

1.3 The Case for Design-Led Innovation

Design is not merely about aesthetics or form; it is a strategic tool for innovation, particularly in safety-critical industries where human lives and well-being are at stake. In such contexts, design serves as a bridge between technical performance and human experience, enabling products to function effectively while accommodating the complexities of real-world use (Hall et al., 2019). Traditional engineering approaches often emphasize conformance to standards and technical specifications, yet they may overlook the dynamic and unpredictable nature of human interaction with safety equipment (Schonheyder, 2019). Empathic design methods and user-centered frameworks offer a powerful counterbalance to this limitation. These approaches prioritize understanding users' emotional, cognitive, and physical needs, thereby fostering designs that are not only functional but also intuitive and inclusive (Koupric and Visser, 2009). Empathy in design is increasingly recognized as a critical competency, allowing designers to "*step into and out of the user's life*" to better grasp the context and consequences of product use (Jiancaro, 2018). This is especially relevant in the development of PPE in general, where discomfort, poor fit, and limited accessibility can undermine safety and discourage consistent use (Dempsey, 2023). For jockeys, this might involve anticipating how fatigue affects vest adjustment, or how stress influences the ability to remove gear during an accident. These considerations are rarely captured by technical specifications alone.

Interdisciplinary collaboration further enhances the design process by integrating diverse perspectives, from engineering and ergonomics to psychology and healthcare (Foster, 2025). Such collaboration ensures that safety equipment is not developed in isolation but reflects the lived realities of its users and the operational environments in which it is deployed. When designers, researchers, and practitioners work together, they can co-create solutions that are both technically robust and human-centered, leading to safer, more equitable outcomes. This principle is central to the development of PPE in high-risk sports (and not only), where comfort, usability, and accessibility are as important as compliance.

The use of PPE has become a cornerstone of modern sports, providing athletes with essential layers of safety. This broad category includes clothing, footwear, helmets, and other wearables specifically engineered to reduce the likelihood and severity of injuries. Because each sport carries distinct risks and performance demands, PPE must be carefully designed to align with the physical, cultural, and ergonomic needs of its users. Yet, within this landscape of innovation, professional horse racing stands out as a discipline that has struggled to keep pace. Despite its global popularity and economic significance, the safety equipment available to jockeys remains comparatively underdeveloped. These athletes continue to rank among the most vulnerable in sport, often relying on gear that is outdated, inconsistently regulated, or resistant to meaningful innovation. In Australia, the introduction of mandatory safety vests in 1998 marked a turning point, contributing to a reduction in jockey fatalities: however, the effectiveness of these vests has frequently been questioned, sparking debate about whether compliance with existing standards is sufficient to protect riders in such a hazardous profession (Foote et al., 2014).

In sum, design in safety-critical domains must evolve beyond compliance and aesthetics. It must embrace empathy, interdisciplinarity, and strategic foresight to meet the complex demands of modern use. This shift not only improves product performance but also empowers users (primary and secondary alike) to engage with safety equipment in ways that are meaningful, effective, and inclusive.

1.4 Structure of the Book

This book is organized into ten chapters that trace the evolution of safety design from static compliance frameworks toward adaptive, human-centered systems, exploring the interplay between design values and product standards across multiple levels. Each chapter builds upon the previous one, moving from theoretical underpinnings to empirical research, and finally to conceptual and practical proposals for reform. The structure mirrors both the research journey undertaken in the Australian jockeys' safety vests study and the broader transformation required in how safety standards are conceived, tested, and governed.

Chapter 1 – Introduction: Why Design Values Matter in Product Standards

The opening chapter establishes the foundation of the book by questioning how safety has traditionally been defined and regulated. It argues that compliance-based systems often prioritize technical performance over human experience and contextual reality. Using Australian jockeys' safety vests as a case study, it highlights the paradox of current regulatory frameworks, systems designed to protect can, at times, inhibit comfort, inclusivity, and innovation. The chapter concludes by presenting the book's central thesis: that integrating design values such as empathy, adaptability, and usability into safety standards is essential for creating protection that truly serves people.

Chapter 2 – Theoretical Foundations: From HCD to D-UX

This chapter develops the conceptual base for the book. It explores Human-Centered Design (HCD) and User Experience Design (UXD), outlining their value and limitations when applied to high-risk domains. The discussion introduces the *Dependency-based User Experience (D-UX)* framework, which expands the design lens beyond direct users to include co-dependent users (such as medical professionals and stewards) whose interaction with safety equipment is equally critical. The chapter emphasizes that understanding these human interdependencies is fundamental to creating effective, inclusive, and adaptive safety systems.

Chapter 3 – Standards as Design Constraints

Here, the book turns its focus to the complex relationship between standards and design. Standards ensure accountability and consistency, yet they can also restrict creativity and experimentation. This chapter traces the origins and evolution of safety vest regulations, demonstrating how well-intentioned protections became rigid frameworks that resist change. Through the case of the *Descente* jockeys' safety, it illustrates how innovation can be obstructed by narrow certification criteria and outdated testing methods. The chapter concludes by proposing a shift toward participatory, design-informed standards that embrace adaptability and continual revision.

Chapter 4 – Case Study: Innovation and Regulation in Australian Jockey Safety Vests

Anchoring the discussion in empirical evidence, this chapter examines the intersection between regulation, design, and lived experience. Drawing on interviews, focus groups, and field observations, it reveals how current Australian standards influence both the design process and the everyday experience of jockeys and medical staff. Participants describe recurring challenges (e.g., poor fit, restricted mobility, and difficulties during emergencies) that expose a misalignment between technical compliance and real-world safety. The findings underscore the need for iterative design processes that integrate user feedback into regulatory reform.

Chapter 5 – The Evolving Role of the User: Innovation, Safety, and Smart Design

This chapter reconceptualizes the user as an active participant in the creation and evolution of safety technologies. It discusses how collaboration between designers, riders, and regulators fosters innovation and more responsive protective systems. By incorporating emerging technologies (such as smart textiles, wearable sensors, and adaptive materials) safety can evolve beyond static protection toward dynamic, data-informed responsiveness. The chapter highlights the ethical and practical implications of this shift, positioning users not as end-points of design but as co-creators of safer futures.

Chapter 6 – Innovation, Standards, and the Paradox of Protection

Here the author delves into the structural and cultural barriers that hinder innovation within Australian safety vest standards. The chapter explains how outdated assumptions about ergonomics, gender, and user experience contribute to design stagnation. A brief historical overview situates these issues within the broader evolution of equestrian safety. It concludes with practical recommendations for revising the standards: establishing pilot-testing pathways, incorporating ergonomic and gender-based criteria, and enabling modular innovation. These proposals illustrate how regulatory flexibility can coexist with safety assurance.

Chapter 7 – From Compliance to Creativity: Conceptual Proposals for Next-Generation Safety Vests

Building upon previous insights, this chapter translates research findings into design imagination. It outlines conceptual goals and speculative prototypes for next-generation vests that integrate adaptability, smart sensing, and modularity. The chapter presents design scenarios and early explorations that demonstrate how protection can evolve as an interactive, personalized experience. It emphasizes the value of interdisciplinary collaboration between designers, engineers, and policy-makers as a means of transforming concepts into viable safety solutions.

Chapter 8 – Beyond Horse Racing: Lessons for High-Risk PPE

The lessons derived from the equestrian context extend well beyond racing. This chapter examines parallels with other high-risk domains such as construction, emergency services, and healthcare, where similar tensions between compliance and usability persist. It identifies shared challenges (such as rigid procurement practices, limited end-user engagement, and cultural resistance to innovation) and argues for cross-sector knowledge exchange. By positioning jockey safety vests as a microcosm of larger systemic issues, the chapter underscores the universal relevance of design-led reform in protective equipment.

Chapter 9 – Rethinking Standards Through Design

This chapter synthesizes the preceding analyses into a cohesive argument for transforming how standards are conceived and managed. It critiques the static nature of traditional frameworks and presents a vision of standards as "living documents" that evolve through ongoing collaboration, data feedback, and participatory review. The discussion outlines institutional and policy implications, suggesting that embedding design thinking into standardization processes can turn regulation into a driver of continuous improvement rather than a barrier to progress.

Chapter 10 – Conclusion: Designing for Safety, Designing for Change

The final chapter brings together the book's central ideas, reaffirming that safety and human experience are inseparable. It argues that design-led approaches can transform protective systems from rigid compliance tools into adaptive, intelligent frameworks that respond to human needs. The closing sections (*Designing for Change, Dignity, and Future Practice* and *Final Reflections*) emphasize the moral and social dimensions of safety design, calling for a culture that protects not only the body but also the dignity, trust, and agency of the individual. The chapter concludes with a forward-looking vision: a future in which standards, design, and human experience evolve together to create systems that are safer, fairer, and more humane

With this book, readers will gain a deeper understanding of how design values can transform not only products but also the systems that govern their development, regulation, and use. This transformation requires a paradigm shift: from compliance-driven frameworks that prioritize minimum thresholds, to value-driven frameworks that integrate human needs, empathy, and lived experience. By reimagining standards through a design-led lens, we can transform them from static, compliance-driven rules into dynamic tools that shape not only safer and more innovative products but also the systems, regulations, and social priorities that govern their creation, ultimately placing human experience, empathy, and shared values at the heart of technological and societal progress.

Chapter 2.
Theoretical Foundations – From HCD to D-UX

Theories shape how we understand, interpret, and design for safety. They provide the intellectual foundation that links practice to purpose, helping designers move beyond intuition and toward intentional, evidence-based decision-making. This chapter examines the evolution of design theory as it relates to safety, tracing the shift from traditional HCD/UCD frameworks toward a more inclusive and dynamic approach, D-UX. It argues that embedding design values within standards requires not only methodological rigor but also theoretical depth, ensuring that innovation remains grounded in empathy, ethics, and context.

2.1 Introduction: Why Theory Matters in Design for Safety

Building upon the discussion in Chapter 1, which highlighted the importance of design values in shaping product standards, this chapter turns to the theoretical foundations that inform how we understand and practice design for safety. Theory provides the conceptual scaffolding that allows designers to question assumptions, interpret human experience, and frame innovation in ethical and contextual terms.

Designing safety equipment is not merely a technical challenge; it is fundamentally a human one. While engineering principles and regulatory standards provide essential frameworks for ensuring mechanical integrity and compliance, they often fall short in addressing the complexity of human interaction, emotional response, and contextual variability (Fargnoli, 2021). This gap underscores the importance of design theory, which offers a lens through which we can understand how people engage with products, how meaning is constructed through use, and how innovation can be guided by empathy, collaboration, and situated knowledge (Gumusel, 2025).

Design theory encompasses a range of methodologies and conceptual frameworks (from user-centered design and human factors engineering to critical design and value-sensitive design) that help bridge the divide between technical performance and lived experience (Doorn & Hansson, 2015). In safety-critical domains such as sports protective equipment, healthcare devices, or industrial machinery, this theoretical grounding becomes indispensable. It enables designers to anticipate not only physical hazards but also cognitive, ergonomic, and emotional challenges that users may encounter (Fargnoli, 2021).

This chapter lays the theoretical groundwork for the book's central argument: that design values must be embedded in product standards, and that doing so requires a shift from traditional user-centered models to a more inclusive and dynamic framework, D-UX (Giusti Gestri, 2019). Unlike conventional user-centered approaches, which focus primarily on direct users, D-UX (see Figure 2) incorporates the roles of secondary and tertiary users, such as paramedics, trainers, and equipment managers.

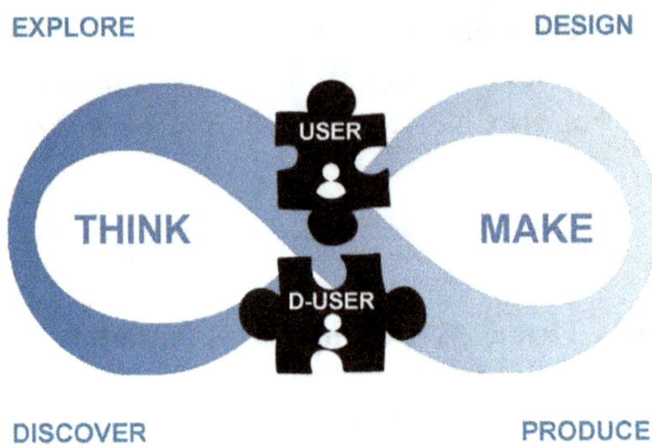

Figure 2. The Dependency-based User Experience (D-UX) framework (Giusti Gestri, 2019)

These actors are integral to a product's lifecycle yet are typically absent from design and standardization frameworks. By adopting D-UX, design evaluation becomes more holistic, accounting for the interdependencies between users, contexts, and systems of use. Hence, embedding D-UX into the discourse on standardization calls for inclusive and participatory design processes that recognize these interdependencies. Participatory design, grounded in democratic and value-centered principles, emphasizes co-designing with users to ensure that their values and lived realities are inscribed into the product itself (Van der Velden & Mörtberg, 2015). This challenges the notion that standards are neutral or universally applicable, arguing instead for adaptive frameworks that reflect diverse user realities and promote equitable access to safety and performance.

D-UX builds upon established HCD and UXD principles but expands their scope to include secondary and tertiary users (such as medical professionals, trainers, and emergency responders) whose interactions with safety equipment are essential yet often overlooked in design and standardization processes (Giusti Gestri, 2024). Hence, the D-UX framework, developed by the author, builds on contemporary research in experience design while offering a novel approach to understanding user experience in product interaction. In particular, it draws on *who* highlight the dynamic, context-dependent interplay between user, system, and environment, emphasizing that user experience emerges through interactions within specific situational and cultural contexts (Berni & Borgianni, 2017). By integrating these insights, the D-UX framework extends traditional models by explicitly mapping dependencies between users, tasks, and contextual factors, providing a structured method for capturing how experiences unfold in real-world settings. Unlike previous approaches, which often treat user experience as a static or isolated phenomenon, the D-UX framework positions experience as an emergent property shaped by interrelated elements, enabling designers to anticipate how design decisions affect both usability and subjective experience in diverse scenarios.

Consequently, the D-UX framework recognizes that safety equipment does not operate in isolation but within a network of dependencies, where each user's role and experience contribute to the product's overall effectiveness. By integrating this perspective into the design and evaluation of standards, we move toward a more holistic, equitable, and responsive approach to safety innovation. Hence, theory matters because it equips designers with the tools to think critically, act ethically, and innovate responsibly. It transforms design from a reactive process of compliance into a proactive strategy for human-centered safety.

2.2 User-Centered Design (UCD): Foundations and Limitations

The concept of HCD, sometimes interchangeably referred to as User-Centered Design (UCD), originates from the disciplines of human factors engineering and ergonomics, which emerged prominently during the mid-twentieth century. During World War II, the complexity of machines and systems necessitated greater attention to the interaction between humans and technology, as poorly designed interfaces often resulted in errors with potentially catastrophic consequences (Norman, 2013). This early focus on optimizing the *fit* between humans and machines laid the groundwork for what would later evolve into HCD.

In the 1980s, Donald Norman and colleagues in cognitive science and design studies articulated a systematic framework for designing with the user at the center. Norman (2013) also critically demonstrated how usability failures are not attributable to human error, but to inadequate design. His arguments reoriented the field of design toward the primacy of human needs, cognition, and behavior.

By the late 20th and early 21st century, the HCD paradigm expanded significantly. With the advent of personal computing, the internet, and mobile technologies, HCD became a cornerstone of digital interface design. Simultaneously, organizations such as *IDEO* and academic institutions like *Stanford's d.school* popularized *design thinking*, embedding HCD within frameworks of innovation, entrepreneurship, and social problem-solving (Brown, 2009). Today, HCD is applied beyond product and interface design to domains such as healthcare, policy-making, and education, where human experiences are integral to successful outcomes. Along with that, the centrality of HCD in product development lies in its capacity to align technological affordances with human values, practices, and contexts of use. From a product development perspective, the following contributions are particularly salient:

Enhancement of Usability and Adoption: HCD methodologies produce products that are intuitive, efficient, and aligned with users' cognitive models, thereby reducing errors and training costs (Giacomin, 2014).

Fostering Innovation through Empathy: By systematically incorporating empathy into design processes, HCD facilitates the discovery of latent needs that users themselves may not articulate, thus enabling innovation beyond incremental improvement (Brown & Katz, 2011).

Risk Mitigation and Cost Efficiency: Early-stage prototyping and iterative testing with users allow design teams to identify failures prior to large-scale deployment, decreasing the financial and reputational risks associated with unsuccessful products.

Inclusivity and Accessibility: HCD provides a framework for incorporating diverse user perspectives, thereby contributing to the development of products that serve broader populations, including marginalized groups.

In an increasingly competitive global economy, the integration of HCD principles enhances not only product adoption but also long-term user satisfaction and brand loyalty. Notably, the theoretical underpinnings of HCD rest on a set of principles codified in standards such as ISO 9241-210:2019 (ISO, 2019), which defines HCD as "*an approach to systems design and development that aims to make interactive systems more usable by focusing on the use of the system and applying human factors and usability knowledge*." The foundations of HCD can be understood through five interrelated principles:

Empathy and User Understanding: Developing a comprehensive and situated understanding of users' goals, practices, emotions, and cultural contexts.

Participatory Engagement: Actively involving users in co-design processes to ensure relevance and legitimacy of design outcomes.

Iterative Development: Employing cycles of prototyping, testing, and refinement informed by user feedback.

Systemic Perspective: Accounting for the broader socio-technical environment in which products and services operate.

Ethical Orientation: Recognizing design as an inherently ethical practice with social, cultural, and ecological implications.

These principles position HCD as a methodological bridge between engineering-driven functionality and human-centered meaning-making. While HCD offers substantial advantages, it is not without theoretical and practical limitations. HCD, while widely praised for its empathetic and participatory approach, has its limitations. Basing design processes on current user practices and preferences can lead to path dependence and incremental changes, unintentionally restricting radical or disruptive innovation. As Verganti (2009) notes, transformative breakthroughs often come from vision-driven strategies that foresee new meanings rather than just addressing existing needs. Additionally, the resource-intensive nature of HCD (due to its repeated cycles and stakeholder involvement) can pose significant challenges for organizations with tight budgets or time constraints. The method also risks bias and representational issues when user groups are unrepresentative, potentially perpetuating systemic inequalities or ignoring wider demographic diversity.

Despite its focus on empathy, HCD may conceal ethical blind spots, failing to fully consider issues like data privacy, sustainability, or unintended societal impacts. Finally, in some institutional settings, HCD is used superficially as a rhetorical device or *buzzword*, without genuine integration into design practice, which diminishes its transformative potential. Acknowledging these limitations is essential for situating HCD within a pluralistic design landscape, where it must be complemented by other paradigms such as technology-driven innovation, critical design, and sustainability-focused design.

Considering the UCD limitations when applied to safety-critical and multi-user systems, in the case of jockeys' safety vests, the primary users (jockeys) are not the only ones whose needs matter. Medical professionals, who interact with the vest during emergencies, are co-dependent users whose experience is equally vital. Traditional

UCD frameworks do not adequately account for these layered relationships. Thus, the involvement of both end users (such as athletes) and co-dependent parties (medical professionals, regulators, safety officers) is fundamental for successful product development in sectors tightly regulated for safety. Through ethnographic research, this chapter explores how empathy and co-design principles not only result in more protective equipment but also improve comfort and acceptance. Practical guides and case insights support implementation.

Designing with empathy is vital for successful user-centered innovation. This chapter reviews design thinking frameworks and the role of direct user engagement, emphasizing iterative feedback and continuous improvement. Real-world examples demonstrate how empathetic approaches yield better outcomes in risk management and user satisfaction.

2.3 User Experience Design (UXD): Beyond Usability

While UCD and UXD have transformed design practice, both frameworks often focus on primary users, the individuals directly interacting with a product. Yet in many real-world contexts, secondary and tertiary users are just as important. UXD represents an evolution of UCD by extending the scope of inquiry beyond functionality and usability toward the experiential, affective, and symbolic dimensions of product interaction. While UCD primarily emphasizes task efficiency, error reduction, and usability metrics (Norman, 2013; ISO, 2019), UXD acknowledges that human engagement with artifacts is simultaneously cognitive, emotional, and socio-cultural in nature (Hassenzahl & Tractinsky, 2006). UXD emerged as a distinct paradigm in the late 1990s and early 2000s, coinciding with the proliferation of consumer-facing digital technologies. Scholars and practitioners recognized that competitive advantage no longer resided solely in technical performance but increasingly in the quality of user experiences (Garrett, 2011). Thus, UXD integrates aesthetic design, affective computing, and symbolic communication into the design process, shifting emphasis from "*can users use it?*" to "*how do users feel when they use it, and what does it mean to them?*"

Consequently, UXD emphasizes three interrelated domains that extend beyond usability:

- **Emotional Experience**: Products evoke emotions ranging from trust and delight to frustration and anxiety. These affective responses shape long-term acceptance and loyalty (Norman, 2004).

- **Aesthetic Experience**: Visual form, materiality, and sensory interaction contribute significantly to product evaluation, with aesthetics often influencing perceptions of usability itself (Tractinsky et al., 2000).
- **Symbolic and Identity Functions**: Products embody meanings and values, enabling users to express identity, status, and cultural affiliation (McCarthy & Wright, 2004).

By addressing these dimensions, UXD provides a holistic understanding of user–artifact relationships within broader life contexts. Hence, its relevance becomes evident when examining specialized products such as safety vests for jockeys. While usability considerations focus on protection, fit, and removal, UXD highlights additional dimensions:

- **Comfort and Restriction**: Jockeys' perception of vests as uncomfortable or restrictive reflects not only physical constraints but also an affective dissonance between safety requirements and embodied freedom of movement.
- **Emotional Resonance in Emergencies**: Medical staff's struggle with vest removal during accidents reveals the emotionally charged context in which product interaction occurs, where stress, urgency, and empathy shape the user experience.
- **Aesthetic and Motivational Aspects**: The perception of vests as uninspiring highlights how product design communicates value and identity, particularly in high-performance environments where equipment is integral to professional image.

Through this lens, UXD demonstrates how safety products are not merely protective devices, but experiential artifacts embedded in the practices, identities, and emotions of their users. Despite its expanded scope, UXD remains constrained by certain underlying assumptions. First, UXD tends to adopt an individual-centric focus. Its methods and evaluative frameworks predominantly privilege the subjective experiences of single users, often at the expense of recognizing the interdependencies that emerge among multiple stakeholders within complex socio-technical systems (Forlizzi & Battarbee, 2004). This limitation is particularly evident in collaborative or high-risk contexts, where user experiences are shaped not in isolation but through interactions with others, technologies, and institutions.

Second, UXD has been critiqued for exhibiting contextual and systemic blind spots. While it effectively addresses immediate interactions between individuals and products, it does not always adequately account for the broader networks within which those interactions occur. In the case of safety equipment, for example, design considerations extend beyond the artifact itself to encompass regulatory standards, institutional protocols, and the coordinated practices of teams operating under conditions of urgency. UXD's primary emphasis on the individual user may obscure these systemic dynamics, thereby limiting its capacity to address the full spectrum of design challenges in such environments.

Finally, the methodological foundations of UXD face inherent challenges. Experience, by its very nature, is subjective, fluid, and multidimensional. Capturing, measuring, and analyzing this complexity within design research has proven difficult, often resulting in inconsistencies across studies and practical applications (Law et al., 2009). These methodological uncertainties raise questions about the reliability and comparability of UX research, particularly when design decisions must be justified to stakeholders beyond the design team.

Acknowledging these limitations underscores the necessity of integrating UXD with broader systemic and multi-stakeholder approaches. This is especially crucial in domains where safety, collaboration, and regulation converge, as the design of products and systems in such contexts must attend not only to individual experiences but also to the interwoven social, institutional, and technical structures that shape them.

Notably, UXD expands on UCD by focusing not just on usability, but on the emotional, aesthetic, and symbolic dimensions of product interaction. It considers how products make users feel, how they fit into daily life, and how they express identity and values. In the context of safety vests, UXD helps us understand why jockeys perceive current designs as uncomfortable, restrictive, and uninspiring. It also explains why medical staff struggle with vest removal during emergencies, an experience that is not just functional but emotionally charged. Yet, UXD still centers on the individual user. It does not fully address the interdependencies between users, nor the systemic nature of safety equipment in high-risk environments. Instead, the concept of D-UX extends UXD to account for these interdependencies (Giusti Gestri, 2019). In the case of Australian jockey safety vests, the primary users are jockeys, but the secondary users include paramedics, trainers, equipment handlers, and even regulatory authorities. Their needs, perspectives, and constraints influence the design's overall effectiveness.

For example:

- Paramedics must be able to remove vests quickly in emergencies.
- Medical professionals need to have the jockeys' medical histories available
- Trainers and staff must ensure vests are properly fitted and maintained.
- Regulators must balance technical safety thresholds with real-world feasibility.

Thus, D-UX builds on UCD and UXD but adds a critical layer: the recognition of co-dependent users whose roles are essential to the product's function and success. In the case of Australian jockeys' safety vests:

- Primary users are jockeys who wear the vest.
- Co-dependent users are medical professionals who must interact with the vest during emergencies.
- Regulatory bodies and designers are stakeholders whose decisions shape the vest's design and standards.

D-UX proposes that all these actors must be considered in the design process. Their needs, constraints, and interactions form Forlizzi's (2007) product ecology (which conceptualizes products not as isolated artifacts but as entities situated within social, emotional, and functional contexts), a network of relationships that determine how the product is used, perceived, and valued. From this viewpoint, the value of a design cannot be understood solely in terms of individual usability but must also be evaluated in relation to the broader ecosystem of practices, relationships, and meanings in which it is embedded. Another important foundation is the theory of design-driven innovation, which emphasizes that meaningful innovation emerges not simply from enhancing functional performance but from reinterpreting and reshaping the symbolic meanings of products (Verganti, 2009). This perspective encourages designers to look beyond incremental improvements and instead challenge established paradigms, thereby positioning design as a driver of cultural and market transformation.

Complementing these approaches, the tradition of empathic design underscores the necessity of gaining deep insight into users' lived experiences (Leonard & Rayport, 1997). By attending to subtle, often unarticulated aspects of everyday life, empathic design provides a pathway for creating products that resonate emotionally and contextually with users, thereby fostering more meaningful interactions. Besides, institutional theory highlights the role of cultural and organizational norms in shaping design acceptance. According to this perspective, products do not gain legitimacy solely through functional performance or user satisfaction, but also through their alignment with prevailing institutional logics and expectations (DiMaggio & Powell, 1983). This insight reminds designers that innovation must not only serve users but also navigate the regulatory, cultural, and organizational environments in which products are situated.

Taken together, these theoretical traditions converge within D-UX to support an approach to design that is at once user-centered, contextually grounded, and strategically oriented. By integrating ecological, cultural, empathic, and institutional dimensions, D-UX moves beyond narrow definitions of usability or user experience to embrace design as a complex, multi-level practice of innovation and meaning-making. Traditional standards, focused narrowly on impact resistance or durability, often neglect these broader contexts. D-UX argues for an ecosystem perspective, where products are designed and evaluated as part of a network of human and organizational interactions.

2.6 Integrating design values into product standards

This chapter has traced the intellectual trajectory from UCD and UXD, and subsequently introduced D-UX as a necessary advancement for contexts where safety, collaboration, and systemic complexity converge. While UCD established a foundation by emphasizing usability and functional alignment with user needs (Norman, 1988; ISO, 2019), UXD broadened the scope to encompass affective, aesthetic, and symbolic dimensions of interaction (Hassenzahl & Tractinsky, 2006; McCarthy & Wright, 2004). D-UX extends these traditions further by embedding design within socio-technical systems, recognizing that safety-critical products operate not only as individual artifacts but also as components of broader institutional, cultural, and regulatory frameworks (Forlizzi, 2007; DiMaggio & Powell, 1983).

The author has outlined three major design frameworks (UCD, UXD, and D-UX) that together offer a robust theoretical foundation for rethinking product standards. Their integration suggests a pathway forward:

1) **UCD** ensures usability and human-centered efficiency.
2) **UXD** expands the scope to include emotional, social, and experiential dimensions.
3) **D-UX** embeds interdependencies, recognizing that products exist in complex ecosystems of use.

Collectively, these approaches create a holistic framework for embedding design values into standards. Rather than constraining innovation, such an approach uses standards as platforms for inclusive, empathic, and participatory design. Consequently, the subsequent chapters operationalize this framework through an in-depth case study of jockeys' safety vests. By examining this highly specialized form of PPE, the discussion will illustrate how design values can be translated into formal standards, how innovation can be fostered within restrictive regulatory environments, and how products can be reconceptualized to address the needs of both primary users (jockeys) and co-dependent users (such as medical staff, trainers, and regulatory authorities).

This dual focus underscores the necessity of moving beyond a strictly individual-centric perspective toward an approach that recognizes the interdependencies shaping design outcomes in high-risk sporting contexts (Forlizzi & Battarbee, 2004). A central tension explored in this inquiry is the relationship between innovation and regulation. In the field of sports PPE, regulatory standards function as essential safeguards for user safety and legitimacy, yet they often constrain the adoption of radical design transformations (Fleck & Howells, 2001; National Academies of Sciences, Engineering, and Medicine, 2023). Consequently, innovation in this sector frequently emerges in incremental forms (minor adjustments to fit, comfort, or usability) rather than in the form of disruptive reimagining of protective equipment.

While incremental innovation tends to align more easily with regulatory requirements and market acceptance (Henderson & Clark, 1990), it may also limit the capacity to address persistent shortcomings experienced by users. Conversely, radical transformation has the potential to redefine both the function and meaning of safety equipment (Verganti, 2009), but faces significant obstacles in terms of certification, compliance, and stakeholder trust (Ashford & Hall, 2011).

Therefore, this chapter provides an analytical framework for evaluating the trade-offs between incremental and radical innovation. By considering their implications for user safety, regulatory compliance, and market adoption, it seeks to illuminate how innovation pathways can be responsibly navigated in safety-critical product domains. The analysis adopts a qualitative approach grounded in documented case studies, allowing for a nuanced exploration of the tensions, negotiations, and opportunities that arise when design values intersect with institutional structures and lived user experiences (Yin, 2018). Innovation in sports PPE often comes into conflict with product standards, which may slow or restrict adoption of radical design.

Summarizing, integrating design values into product standards requires a dual commitment: to rigor and to empathy. The theoretical frameworks discussed in this chapter demonstrate that safety cannot be ensured through compliance alone: it must be cultivated through understanding, participation, and responsiveness to real human contexts. The D-UX model offers a means of reconciling the demands of regulation with the realities of lived experience, framing safety as a shared, evolving responsibility among all stakeholders.

Thus, the next chapter translates these theoretical insights into practice. By examining the case of jockeys' safety vests, it reveals how the principles of D-UX can be operationalized within an industry where performance, tradition, and safety intersect. This transition from theory to application marks a crucial step toward redefining how standards are conceived—not as static prescriptions, but as dynamic frameworks that evolve with human needs and technological progress.

Chapter 3.
Standards as Design Constraints

Design and regulation have long existed in a delicate balance. While design seeks to imagine new possibilities, standards seek to preserve consistency, reliability, and safety. Nowhere is this balance more evident than in safety-critical contexts such as professional horse racing, where protective equipment must meet stringent technical specifications while accommodating the complex physical and psychological needs of athletes. This chapter builds on the theoretical framework introduced in Chapter 2 by examining how design values interact with regulatory structures in practice.

Through the case of Australian jockeys' safety vests, the discussion reveals how standards (though essential for safeguarding lives) can inadvertently constrain innovation and user experience. By analyzing the evolution, intent, and application of key Australian standards such as EN 13158 and ARB 1.1998, this chapter investigates how regulatory systems influence not only the physical form of safety equipment but also the broader design culture surrounding it. The aim is to uncover how standards can transition from being mere instruments of compliance to becoming catalysts for human-centered innovation.

3.1 Introduction: The Double-Edged Role of Standards

Product standards play a pivotal role in ensuring safety, consistency, and accountability across industries. In high-risk contexts such as horse racing, they serve as regulatory cornerstones, defining acceptable levels of protection, guiding manufacturing processes, and formalizing the criteria by which products are tested and approved. These frameworks, embodied in documents like EN 13158 and ARB 1.1998, specify requirements for impact absorption, coverage zones, and structural integrity, aiming to protect jockeys from life-threatening injuries.

Yet, the very mechanisms that ensure reliability can also act as constraints on innovation. Once codified, standards tend to resist change: revisions are infrequent, and adaptation often occurs only within narrow technical margins. As a result, the creative and human dimensions of design (e.g., fit, comfort, usability, and emotional engagement) may be sidelined in favor of compliance. This rigidity generates a persistent tension: while meeting standards ensures minimum safety thresholds, it can simultaneously stifle opportunities for user-centered innovation and the integration of emerging materials or technologies.

In the case of jockeys' safety vests, this tension is particularly acute. The existing standards were developed from a predominantly technical perspective, prioritizing impact performance and material durability. However, they rarely account for the lived experiences of the athletes who must perform under extreme physical and psychological pressure. Jockeys, whose work demands agility, balance, and precision at high speeds, often perceive safety vests as cumbersome or restrictive: an indication that the protective design has not fully reconciled functional safety with experiential usability.

This section therefore examines the dual role of standards as both enablers and inhibitors of design progress. By exploring how these frameworks shape, limit, and sometimes inspire innovation in safety vest design, it highlights the need to reimagine standards not as static prescriptions, but as evolving design instruments, ones capable of accommodating both human diversity and technological advancement.

3.2 The Origins and Intent of Safety Vest Standards

Equestrian body protectors are governed by several international standards that specify impact resistance, coverage, and material performance. Among the most globally recognized are EN 13158, ASTM F1937, and ASTM F2681. These frameworks have been instrumental in establishing minimum safety thresholds, yet most have seen few substantive updates since the early 2000s, despite significant advances in materials science, digital fabrication, and wearable technologies (ASTM International, 2023; BETA, 2024).

In Australia, jockey safety vests are still regulated under both ARB 1.1998 and EN 13158, which have been jointly enforced since 1998 (Racing Australia, 2025). These standards were designed to provide broad protection across rider body types and to account for anatomical differences between men and women. However, ARB 1.1998 in particular has not evolved to reflect contemporary ergonomic or physiological research, an oversight that has become increasingly visible as the proportion of female jockeys has risen (Foote et al., 2011, 2014).

Both EN 13158 and ARB 1.1998 codify requirements for impact attenuation, coverage area templates, thermal performance, labeling, and certification. Their stated goal is to reduce the severity of injuries during falls. Yet, as empirical studies and user feedback reveal, these standards are built on narrow anthropometric assumptions. The author's earlier research (Giusti Gestri, 2019, 2020, 2023a,b, 2024) documents widespread reports of poor fit, particularly among female riders, with vests that fail to accommodate variations in torso shape, chest size, and body proportions. This evidence underscores a fundamental design gap: while standards ensure mechanical safety, they insufficiently address the diversity of real bodies and performance contexts.

Although EN 13158:2018 nominally requires testing across children, youths, and adults "of either sex" and in "at least seven sizes," these parameters do not guarantee inclusion of extreme or atypical morphologies. Studies of New Zealand jockeys (Legg et al., 2023) highlight substantial variation in strength, balance, and physique across gender and experience level, differences that directly affect how safety vests perform in use. Because compliance with approved standards is mandatory in most jurisdictions, including under Racing Australia rules, these design templates effectively become regulatory constraints rather than flexible guidelines.

The Australian Racing Board (ARB) introduced its own standard, *ARB 1.1998*, to mandate safety vest use in all races and training sessions. Adapted from the earlier EN 13158 standard (first issued in 1986), it was tailored for local climatic and environmental conditions, such as high temperatures and humidity. This adaptation illustrates that standards are not neutral or universal: they are socially and geographically situated artefacts, reflecting assumptions about context, technology, and the bodies they are designed to protect (Busch, 2011). Since the late 1990s, however, the racing industry has transformed, and the following should be considered:

- The number of female jockeys has grown substantially.
- Smart materials and wearable sensors have matured.
- Climate variation and track conditions have diversified.
- Medical professionals now play more active roles in on-track emergency care.

Despite these shifts, the standards have remained largely static. They do not account for gender-specific ergonomics, modular systems, or sensor integration. Nor do they recognize the needs of secondary users, paramedics, safety officers, and equipment managers, whose interactions with the vest affect both safety and post-incident care.

The ARB 1.1998 standard prescribes specific materials and configurations: perforated foam strips of varying thickness encased in polyester mesh, secured with adjustable straps or Velcro®. These requirements were designed to balance impact absorption, durability, and comfort under physical exertion in hot climates. However, by prescribing material composition and construction methods, the standard also limits experimentation with new lightweight composites or adaptive designs. Only two approved vest templates are authorized for both male and female jockeys (Foote et al., 2014, p.30; Giusti Gestri, 2023a, p.8), effectively constraining the potential for ergonomic or anthropometric customization. Figure 3 illustrates examples of approved vests, highlighting the narrow design scope permitted within existing regulations.

Figure 3. A few of the Australian jockeys' safety vests that are currently approved for races (author's work)

The enduring reliance on these legacy standards exposes a deeper tension between regulatory compliance and design evolution. From a design theory perspective, ARB 1.1998 exemplifies the paradox of standardization: it provides reproducibility and legal clarity, yet risks flattening bodily diversity and overlooking ergonomic nuance. The jockey's body, central to both athletic performance and the economics of racing, becomes simultaneously a subject of protection and a subject of standardization. In this sense, the safety vest functions as more than a protective object; it embodies the intersection of institutional priorities, technological limitations, and lived corporeal experience.

The author's study (2019) with Australian jockeys reinforces this point: riders frequently describe current vests as restrictive, uncomfortable, and poorly fitted, especially for those with non-average builds or in hot, humid conditions. These observations reveal that, while the *intent* of safety vest standards is to reduce injury severity, their *execution* reflects a historically narrow conception of the user, typically male, of average build, and competing in temperate environments.

A design-led reimagining of standards could bridge this gap. By embedding human-centered and dependency-based principles (via the D-UX) into regulatory frameworks, it becomes possible to design protective equipment that reflects *real* anatomical, ergonomic, and contextual diversity. In doing so, standards could evolve from static compliance checklists into dynamic design instruments, frameworks that not only preserve safety but also promote innovation, inclusivity, and adaptability in high-risk sports.

3.3 Innovation, Design, and the Constraints of Standardization

Before examining how standards constrain design innovation, it is essential to understand the broader discourse around innovation itself, how it is defined, operationalized, and contested across fields. Innovation has become not only a buzzword in design but also a central policy goal for governments, industries, and international organizations. Yet, its meaning and implementation vary widely, particularly when mediated through systems of regulation and standardization.

Innovation has become a much-invoked concept in the public sector over the last two decades (Hartley et al., 2013; Osborne & Brown, 2013). Faced with increasingly complex social, economic, and environmental challenges, alongside enduring financial constraints, governments have turned to innovation as a guiding principle for reimagining service delivery and policy design (Lewis et al., 2020). While the term carries strong normative appeal (signifying progress, adaptability, and creativity) its application often collides with the procedural stability and predictability of bureaucratic institutions. Despite these tensions, innovation has gained institutional legitimacy, illustrated by the rise of dedicated Public Sector Innovation (PSI) labs worldwide, such as the OECD's Observatory of Public Sector Innovation (OPSI) established in 2014 (Lewis et al., 2020).

Within this landscape, design-led approaches have emerged as key mechanisms for fostering innovation. Often framed under the banner of design thinking, these methods draw on the iterative, empathetic, and problem-reframing practices of professional design disciplines. Dorst (2011) famously defines design thinking as "the parallel creation of a thing (object, service, system) and its way of working" (p. 525), emphasizing the intertwined nature of form, function, and meaning. By situating creativity within structured experimentation, design-led approaches offer a bridge between institutional rigidity and adaptive learning. This development also signals what Chen et al. (2016) term a "social turn" in design, a shift from designing products to designing systems, relationships, and social change. Influenced by participatory thinkers such as Papanek and Manzini, this movement situates design as a tool for collective problem-solving, embedding empathy, co-creation, and contextual awareness at the heart of practice (Bjögvinsson et al., 2012). In theory, such methods challenge linear, compliance-oriented approaches to governance. Yet, in practice, their impact is often constrained by the very frameworks meant to ensure safety and fairness: standards and regulations.

One of the central paradoxes of standardization lies here. While standards guarantee safety, interoperability, and accountability, they can also inadvertently inhibit innovation. This tension is particularly visible in the field of sports PPE, where standards not only establish performance thresholds but often prescribe materials, construction methods, and testing protocols. The Australian context provides a telling example. The ARB 1.1998 and EN 13158 standards governing jockey safety vests define impact resistance, coverage areas, and structural integrity, ensuring a baseline of protection. However, by codifying specific materials and testing procedures, these standards inadvertently narrow the scope for design experimentation. Promising innovations, such as lightweight composites, smart textiles, and body-mapped protective systems, are often excluded, not for lack of protective efficacy, but because existing testing regimes cannot accommodate them without extensive and costly revision (ASTM International, 2023; BETA, 2024).

Manufacturers thus face a dilemma: to innovate and risk non-compliance, or to comply and accept incremental rather than transformative change. This environment fosters what Allen and Sriram (2000) call a "compliance-first design culture," where the pursuit of certification overshadows user experience, inclusivity, and aesthetic or ergonomic refinement. Consequently, many safety vests remain rigid, bulky, and poorly fitted, prioritizing impact resistance over comfort and usability (National Academies of Sciences, Engineering, and Medicine, 2023).

The restrictive effects of standardization extend beyond materials and form. Standards tend to privilege test performance over lived experience (Aryana, Osvalder, & Borell, 2024), leading to the exclusion of features such as modular padding, magnetic fasteners, or medical access points—features that could improve both usability and emergency response (Babik, Alper, & Downey, 2023). Likewise, gender inclusivity remains an unresolved issue. Current sizing systems rarely accommodate female body morphology, reinforcing inequities and diminishing protective efficacy for a growing cohort of female jockeys (Foote et al., 2011; Janson, 2023). This dynamic is not confined to equestrian sports. Across PPE industries, standards tend to reward incremental innovation (slightly lighter foams, marginally improved ventilation) while discouraging radical redesigns that would require revising entrenched regulatory frameworks. Standards that initially aimed to protect users can thus become barriers to progress, locking in design templates and restricting the evolution of practice.

Viewed critically, standards are not neutral technical artefacts but cultural and political constructs. They embody negotiated compromises between institutional logics, technological affordances, and lived bodily realities (Busch, 2011; Anderson, 2010). In this light, the jockey safety vest becomes a site where governance, risk management, and design culture intersect, a physical manifestation of the balance between innovation and compliance.

From a design-led perspective, the goal is not to reject standards but to reimagine them as adaptable, human-centered frameworks. Incorporating principles of ergonomics, diversity, and contextual adaptability could transform standards from static compliance tools into dynamic instruments that guide and inspire innovation. Ultimately, the evolution of jockey safety vests (and sports PPE more broadly) illustrates the need for a paradigm shift: from compliance-driven systems to value-driven systems that place human experience, diversity, and adaptability at the center of technological and regulatory design. In doing so, standards can move beyond constraint to become catalysts for safer, smarter, and more inclusive forms of innovation.

3.4 Design Innovation vs. Standards Compliance: The Case of the *Descente* jockeys' safety Vest

The Japanese Descente vest (shown in Figure 4), widely praised by jockeys for its superior comfort, flexibility, and overall wearability, exemplifies the tension between user-driven innovation and standards compliance in safety equipment. Its design employed variable-density foams to optimize energy absorption across different impact zones, ergonomic shaping to accommodate dynamic body movement, and modular components that enhanced both fit and ease of donning and removal. Collectively, these features represented a significant departure from the uniform, rigid construction of earlier protective vests. Jockeys consistently reported that the Descente vest reduced perceptions of bulk, allowed greater range of motion, and was more compatible with the intense physical demands of racing (Giusti Gestri, 2019).

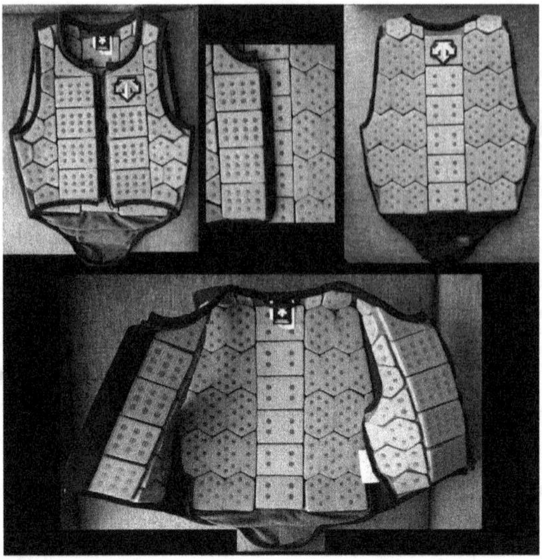

Figure 4. The Japanese Descente jockey's vest (author's work)

Despite its popularity in Japan and its international recognition as a more user-centered design, the vest failed to meet the requirements of the ARB 1.1998 standard in Australia. Specifically, the standard's impact attenuation test protocols demanded uniform thresholds across all panels, limiting the use of graded or multi-density materials; its gap measurement requirements prescribed continuous coverage templates that clashed with modular construction. As a result, the vest was formally disqualified, even though users feedback indicated it was perceived as both safer and more comfortable than compliant models (Foote et al., 2014).

The Descente safety vest, developed in Japan and certified by the Japan Racing Association (JRA), has gained recognition for its lightweight construction, enhanced flexibility, and superior ventilation, which collectively improve jockey comfort and mobility during races. Its recent adoption by the Hong Kong Jockey Club (HKJC), demonstrates confidence in the vest's performance and the robustness of Japanese safety standards (The Hong Kong Jockey Club, 2025, p.25). In contrast, Australia continues to prohibit its use, adhering to outdated standards such as ARB 1.1998 and EN 13158, which have not been significantly revised since the early 2000s (Giusti Gestri, 2019; Giusti Gestri & Barnes, 2020). This regulatory inertia has prompted calls for reform, especially following serious injuries and fatalities among jockeys, with critics arguing that current vests may exacerbate spinal and neck injuries due to their rigidity (Dobbin, 2024). The disparity between countries highlights how product standards can either facilitate or hinder innovation, with Japan and Hong Kong embracing newer, user-informed designs, while Australia remains constrained by legacy frameworks. While intended to guarantee minimum protection, standards such as ARB 1.1998 may inadvertently suppress design innovation by privileging measurable, easily testable performance criteria over holistic user experience and ergonomic benefits. As several scholars have argued, standards both enable and constrain technological trajectories: they stabilize markets and protect users, but they also institutionalize specific assumptions about risk, body morphology, and performance that may lag behind advances in design and material science (Blind, 2016; Giusti Gestri and Barnes, 2020).

More than a decade ago, an Australian study (Foote et al., 2014) aimed to investigate the effectiveness of jockeys' safety vests and trackwork riders in use at that time. Thus, the performance of jockey safety vests was highly dependent on the configuration and properties of their padding materials. They even consider the Descente vest, which utilizes foam pieces of variable density, with softer padding around the neck and arm openings and firmer foam in the chest and back. While this design enhances comfort and flexibility, the report found that it compromises impact protection in critical areas. Impact testing was conducted only on the denser foam sections, presenting the vest's optimal performance. The Descente vest demonstrated marginal results in the flat impactor test, yet considering its minimal padding thickness, the foam material performed reasonably well. However, the vest failed the narrow bar impactor test, largely due to large gaps between padding pieces and the absence of coverage over the shoulders and behind the front zipper. These gaps likely reflect a design prioritizing adjustability and wearer comfort, but they also contribute to reduced energy absorption in high-impact scenarios. Foote et al.'s (2014) comparative analysis reveals that Tipperary and Racesafe vests offer greater circumference coverage and satisfy gap test requirements, aligning more closely with regulatory standards. Notably, the European Standard EN 13158 requires compliance with two distinct impact configurations: the flat impactor test and the narrow bar impactor test. Vests must balance padding thickness and bending stiffness to pass both. Softer or thinner vests like Descente and Tipperary typically fail both tests, while stiffer designs such as the Komperdell Ballistic may pass the impactor test but fail the narrow bar test due to excessive rigidity. The narrow bar test, with its smaller contact area, demands either increased stiffness or greater padding thickness to ensure adequate protection.

In contrast, Australia's ARB Standard 1.1998 only mandates compliance with the flat impactor test, omitting the narrow bar test entirely. This regulatory omission reduces design constraints but raises concerns about the adequacy of protection in real-world racing conditions. The implications of excluding the narrow bar test remain underexplored, and the suitability of the narrow bar impact test for assessing injury risk in horse racing warrants further investigation. The Descente vest's failure to meet gap and impact test requirements under these standards highlights a broader issue: current regulations may inhibit innovation by favoring rigid technical specifications over user-informed design. This tension is particularly evident in Australia, where regulatory conservatism has delayed the adoption of internationally successful designs (e.g., the Descente vest). The author considers it important to note that both safety vest standards (EN 13158 and AR 1.1998) are still in use in Australian thoroughbred racing (Racing Australia, 2025, p.70).

Hence, the controversy surrounding the Descente jockeys' safety vest exemplifies a critical tension in sports PPE innovation: while regulatory frameworks are designed to safeguard athletes, they can inadvertently hinder the adoption of empirically validated, user-preferred designs. This tension is particularly evident when international success and user advocacy clash with entrenched technical standards. To move forward, there is a pressing need for adaptive and reflexive regulatory systems, ones that integrate ethnographic user data, cross-cultural performance evidence, and advancements in material science, rather than relying solely on static, legacy protocols. Such an approach would better align safety standards with the lived realities of athletes and the evolving landscape of protective technologies.

3.5 Toward Design-Informed Standards

For standards to act as enablers rather than inhibitors of innovation, they must evolve beyond narrow, prescriptive frameworks. Increasingly, scholars argue that standards should not only codify minimum safety requirements but also embody the lived realities of use, the diversity of user populations, and the potential of emerging technologies (Blind, 2016; Giusti Gestri, 2020). In equestrian sports, this calls for a design-informed reimagining of protective equipment standards, one that integrates insights from human-centered design, ergonomics, and systems thinking.

A first step involves embedding UCD principles into the standards framework. Comfort, fit, and usability should not be treated as secondary to protection but recognized as critical determinants of both safety performance and regulatory compliance (Norman & Draper, 1986; Giusti Gestri, 2019). When riders experience discomfort, overheating, or restricted mobility, they are less likely to wear protective gear correctly or consistently, compromising its effectiveness. Design-informed standards would also acknowledge co-dependent users (such as medical staff and paramedics) whose ability to remove or manipulate a vest quickly in emergencies forms part of the system of safety itself (Forlizzi & Battarbee, 2004). In this sense, safety vests should be understood not merely as objects of protection but as interfaces within a collaborative ecosystem of care.

Equally important is the need to reform testing protocols. Current standards prioritize laboratory-based evaluations, impact attenuation, gap coverage, and material integrity, under static or idealized conditions. These tests rarely capture the dynamic environments in which jockeys operate such as fluctuating temperatures, high humidity, continuous physical exertion, and repeated mechanical stress. More ecologically valid testing regimes would simulate these real-world contexts, offering a truer measure of protective performance and wearability.

Design flexibility should also be central to future standards. Rigid, template-based criteria constrain the ability of designers to adapt products to different body shapes, genders, and environmental conditions. Modular and ergonomic innovations (e.g., adjustable padding systems or reconfigurable panels) can significantly enhance both fit and comfort. Yet these advances are often excluded under current compliance schemes. The exclusion of the Japanese Descente vest from Australian competition, despite its strong user preference, illustrates how inflexible testing protocols can suppress promising and contextually responsive design solutions (Giusti Gestri, 2019).

Furthermore, as smart technologies become increasingly viable, standards must begin to accommodate and encourage their integration. Wearable sensors, smart fabrics, and biomechanical monitoring systems can provide real-time feedback on impact forces, physiological strain, and vest condition, transforming PPE from passive protection into active safety systems (Dunne, 2010; Giusti Gestri, 2023b). Supporting such innovations requires a shift from prescriptive to performance-based standards, those that assess outcomes rather than mandate fixed forms or materials (Gunningham, 2007; Soltani & Pouypouy, 2019). This approach would preserve safety while enabling creativity and adaptability in design practice.

Finally, the development of standards themselves should adopt a participatory and iterative process. Engaging designers, users, regulators, and medical professionals in the co-creation of safety criteria ensures that standards reflect the practical, embodied, and social dimensions of safety, not just its technical definitions (Giusti Gestri, 2020). Such collaboration aligns with design-led governance principles, transforming standards from static rulebooks into dynamic frameworks for continuous learning and innovation. Overall, design-informed standards offer a pathway toward reconciling the protective and innovative imperatives of sports PPE. By embracing user diversity, real-world performance, and participatory processes, regulatory frameworks can evolve from instruments of compliance into catalysts for inclusive, adaptive, and human-centered safety innovation.

3.6 Rethinking Standards: Toward Participatory and Design-Informed Frameworks

Standards are indispensable for safeguarding athletes, yet their prescriptive nature often constrains innovation and neglects the lived realities of users. Evidence from domains such as aviation and medical devices demonstrates that when human factors, ergonomics, and user trials are integrated into the regulatory process, the resulting standards are not only more robust but also more widely accepted. This participatory approach, involving designers, end-users, and medical professionals, reframes safety as a multidimensional construct, encompassing not only impact resistance but also comfort, mobility, inclusivity, and the capacity for medical intervention.

For equestrian sports, this shift is particularly pressing. Safety vests cannot be reduced to technical metrics or static design templates. Jockeys lived experiences are themselves safety-critical, and emerging technologies (such as smart fabrics, modular padding systems, and wearable sensors) demand flexible regulatory frameworks. By embedding participatory and design-led practices into the Australian standards, such as EN 13158 and ARB 1.1998, regulation can evolve from a compliance-driven constraint into a platform for innovation. In this sense, participatory standards do not dilute safety; they expand it. They acknowledge the body as diverse, situated, and dynamic, while ensuring that rules remain attuned to technological progress and practical use. These reimagining positions standards not as barriers but as enablers of safer, more inclusive, and more innovative protective equipment.

The next chapter builds directly on this conceptual foundation by presenting empirical findings from fieldwork with jockeys and medical professionals, demonstrating how the limitations of current standards manifest in practice and how design values can be operationalized within regulatory frameworks.

Chapter 4.
Case Study: Innovation and Regulation in Australian Jockey Safety Vests

Building on the conceptual arguments developed in the previous chapters, the author now focuses on the case study, which examines how standards operate in practice through the example of safety vests mandated for jockeys in Australian thoroughbred horse racing. This context provides a unique opportunity to investigate the lived consequences of prescriptive regulation, the challenges it poses for design innovation, and the ways in which users, regulators, and manufacturers negotiate the balance between safety and performance.

4.1 Introduction: A Real-World Lens on Design and Standards

Imagine a jockey hurtling down a racetrack at over 60 kilometers per hour, the horse beneath them a powerful, unpredictable force. In these fleeting, high-stakes moments, a safety vest is not merely protective gear: it is a lifeline. Yet, despite rigorous industry standards and decades of design evolution, questions remain: How well do these vests meet the real needs of their users? Are design and regulation aligned with the lived realities of those who depend on them most?

This chapter explores these questions through a case study of safety vests for jockeys in the Australian thoroughbred horse racing industry, drawn from the author's doctoral research. The study situates design within the broader social and regulatory context, highlighting the interplay between human factors, safety standards, and the practical constraints of high-performance sports. Both primary users (jockeys) and co-dependent users (medical professionals) were engaged to capture the full spectrum of experience, from on-track performance to emergency response (Smith, 2018; Giusti Gestri, 2023a,b). Using qualitative methods, including ethnographic observation, semi-structured interviews, and a focus group, the author uncovered significant gaps between the intentions of product standards and the realities of everyday use. These gaps manifest in usability challenges, safety compromises, and missed opportunities for co-design, revealing that regulatory compliance alone does not guarantee effective protection. By examining these lived experiences, this chapter provides critical insights into how safety equipment can be more effectively designed, evaluated, and regulated in high-risk professional contexts.

Through this case study, readers are invited to consider not only what standards prescribe but also how design values are enacted, or overlooked, when products meet real-world conditions. This perspective underscores the necessity of integrating user experience, ergonomic insights, and regulatory frameworks into a cohesive approach to safety design.

4.2 Research Design and Methodology

The author's case study employs a qualitative, design-led approach to explore the lived experiences of professional jockeys and medical professionals in their interactions with a critical piece of personal protective equipment: the safety vest mandated in Australian thoroughbred horse racing. The research is informed by principles of HCD, UCD, and UXD and also product ecology, enabling an examination of how design, context, and user perception converge to shape both the effectiveness and acceptance of safety gear (Giusti Gestri, 2019; Giusti Gestri & Barnes, 2020).

Although not an epidemiological study, the research draws inspiration from methodologies commonly used in health-related contexts, particularly the focus on user interaction within specific cultural and physical environments. The study is positioned as an instrumental case study, designed to generate insights that extend beyond this single context and inform broader considerations in the design and regulation of protective equipment. A dual theoretical framework guided the analysis: HCD/UCD, which foregrounds the experiences and needs of primary users, and the D-UX, a framework developed to account for the roles of co-dependent and secondary users in high-risk settings (Giusti Gestri, 2019).

A purposive sample of twenty participants was recruited, comprising seventeen jockeys (apprentice, professional, and retired) and three medical professionals (paramedics and racecourse doctors). Data collection was multi-modal and iterative, incorporating semi-structured interviews to explore vest use, comfort, safety perceptions, and emergency scenarios; a focus group with apprentice and retired jockeys to examine shared experiences and design preferences; and field observations at racecourses and training facilities to assess vest interaction in authentic contexts (Giusti Gestri, 2023a, b).

The triangulation of interviews, focus groups, and observations was essential for enhancing the validity, credibility, and richness of the findings. By integrating multiple perspectives and sources of evidence, the study captures the complex interplay between regulatory standards, user experience, and the practical realities of safety vest use in thoroughbred racing (Giusti Gestri & Barnes, 2020). This approach provides a detailed, contextually grounded understanding of how design, regulation, and lived experience intersect in shaping both the safety and usability of protective equipment.

4.2.1 Interview Protocols

The qualitative component of this study relied on semi-structured interviews with jockeys and medical professionals to explore their experiences, perceptions, and needs related to safety vests. The interview questions were designed to capture insights into comfort, usability, inclusivity, emergency access, and the interplay between design and safety outcomes. For jockeys, questions focused on the fit, mobility, and overall experience of wearing safety vests during training and racing, including specific challenges faced by female riders or those with smaller or non-standard body types. Medical professionals were asked about their experiences with the vest (and its removal) and intervention during emergencies, their perceptions of vest design, and their suggestions for improving both safety and usability.

Ethical considerations were central to the research design. All participants provided informed consent prior to the interviews, with explicit explanations regarding the purpose of the study, voluntary participation, and the right to withdraw at any time without consequence. Confidentiality was rigorously maintained, with personal identifiers removed from transcripts and data stored securely in accordance with institutional ethics guidelines. The study received approval from a subcommittee of Swinburne University of Technology's Human Research Ethics Committee (SHESC), ensuring that both the welfare of participants and the integrity of the research were upheld. The participants provided their written informed consent to participate in this study. This approach enabled a rich, nuanced understanding of the lived experiences of jockeys and medical professionals, providing critical evidence to inform both the critique of existing standards and the development of design-led, user-centered proposals.

4.2.2 Focus Group and Observation Protocols

In addition to individual semi structured interviews, the study incorporated focus groups and direct field observations to capture collective perspectives and contextual behaviors related to safety vests. Focus groups brought together jockeys and, separately, medical professionals to facilitate discussion of shared experiences, challenges, and potential improvements in vest design. For jockeys, focus groups explored issues of comfort, mobility, fit, and aesthetic preferences, while allowing participants to compare and critique different vest models. Medical professional groups focused on emergency access, ease of removal, and potential design modifications to improve intervention efficiency. The interactive format encouraged participants to reflect on co-dependent interactions, generating insights that might not emerge in one-on-one interviews.

Field observations complemented these discussions by providing real-world data on how vests are worn, adjusted, and modified during training and racing, and how medical staff interact with them during practice scenarios and emergencies. Observations focused on posture, range of motion, emergency removal procedures, and informal modifications or adaptations made by jockeys to improve fit or comfort. The data collected allowed for triangulation with interview and focus group findings, enhancing the validity of the study.

Ethical considerations were maintained throughout. All participants received clear information regarding the purpose of observations and focus groups, and informed consent was obtained prior to participation. Observations were conducted unobtrusively to minimize disruption and protect privacy, with no identifying information recorded. This combination of group discussion and observational data provided a comprehensive understanding of user experience, highlighting gaps between formal standards and real-world practices, and informing the design-led recommendations presented in subsequent chapters.

The study employed a combination of interviews, focus groups, and field observations to capture a comprehensive understanding of user experiences with jockeys' safety vests. Thus, Table 1 provides a detailed overview of each research stage, including the participants involved, the primary focus of inquiry, and the ethical measures taken to ensure informed consent, confidentiality, and participant welfare.

Table 1. Data collection summary

Activity	Participants	Purpose / Focus	Ethical Considerations
Individual semi-structured interviews	Jockeys (male and female); medical professionals	Explore lived experiences with safety vests, including comfort, fit, mobility, emergency access, and design suggestions	Informed consent obtained; confidentiality maintained; audio recorded; participants could withdraw at any time; approval from SHESC
Focus Groups	Jockeys (male and female)	Facilitate discussion of shared experiences, challenges, and design improvements; capture co-dependent interactions	Informed consent obtained; anonymity maintained; discussions recorded only with permission; participants could withdraw
Field Observations	Jockeys during training/racing; medical staff during practice scenarios	Observe real-world use of safety vests, modifications, mobility, emergency removal procedures, and interaction with users	Observations conducted unobtrusively; no identifying information recorded; participants informed and consented; privacy protected

Collectively, these methods not only captured the lived experiences and practical challenges of users but also directly informed the design-led proposals and recommendations presented in later chapters, ensuring that every conceptual innovation is grounded in real-world evidence and human-centered insights.

4.3 Key Findings Summary

Individual semi-structured interviews and field observations revealed a consistent pattern of dissatisfaction among jockeys regarding the comfort, fit, and usability of current safety vests. Riders described the vests as bulky, restrictive, and poorly fitted, with female jockeys particularly affected by designs that fail to accommodate body diversity and gender-specific morphology (Dorries, 2025; Foote et al., 2011, 2014; Giusti Gestri, 2019, 2023 a,b, 2024). The rigid foam panels and limited adjustability constrained movement resulting in both physical discomfort and reduced performance.

To understand the recommendations presented in this book, it is useful to consider the evolution of jockey riding styles. In the early twentieth century, American jockey Tod Sloan developed a novel posture, originally called the "monkey crouch," which, though unconventional at the time, laid the foundation for modern racing techniques. By shortening his stirrups and positioning himself high in the saddle near the horse's neck, Sloan aimed to reduce drag and enhance the horse's efficiency. As show in Figure 5, this approach eventually evolved into what is now referred to as the "*Martini Glass*" riding style.

Figure 5. Jockeys' riding style, better known as the Marini glass (author's work)

For the jockeys, posture is pivotal. The Martini Glass position allows riders to act as a dynamic support system, with their legs functioning like bungee cords to absorb motion while minimizing interference with the horse's stride. During the final 400 meters of a race, most jockeys adopt this posture to let horses move freely, potentially increasing pace by five to seven percent. Motion tracking studies show that the rider's reduced movement compared to the horse contributes to this efficiency (Pfau et al., 2009). Achieving and maintaining this posture requires intensive training focused on core strength, lower-body conditioning, and balance, along with upper-body strength to control the horse effectively. As confirmed by participants, despite its performance advantages and widespread adoption, maintaining the Martini Glass posture remains challenging, particularly due to the constraints of safety vests and the risk of helmet-related injuries.

Although jockeys acknowledged the protective function of the vests, many expressed concerns that the equipment designed to safeguard them may, paradoxically, increase risk by limiting agility during falls. Several noted that vest rigidity interfered with their ability to roll or reposition safely after impact, while others questioned whether existing padding configurations genuinely mitigated injury severity (Giusti Gestri, 2023a,b). These accounts highlight an enduring gender and ergonomic bias within both product development and the regulatory standards that guide it (Anderson, 2010; Busch, 2013).

Medical professionals echoed these concerns, emphasizing that current safety vests impede emergency response and treatment. The lack of intuitive openings or modular components delayed medical access to injured areas, particularly during time-sensitive interventions (Babik, Alper, & Downey, 2023). Paramedics and on-track doctors reported that vest rigidity and uniform fastenings obstructed rapid assessment or stabilization, reinforcing the perception that design priorities favor certification over clinical practicality (Fargnoli, 2021). These professionals also observed that medical expertise is largely absent from standards development, despite their critical role as co-dependent users whose insights could improve design performance and safety outcomes. They also reported how, in case of jockeys' falls laying on the ground, the safety vets' zips may impact their chins or contribute to extra injuries. Their feedback suggests significant opportunities for integrating smart technologies, such as embedded sensors for real-time physiological monitoring and detachable panels to expedite removal during emergencies (Giusti Gestri, 2023 a, b; Dunne, 2010).

Field observations further reinforced these findings. Jockeys frequently modified or customized their vests, adjusting straps, trimming foam layers, or choosing lighter models to balance protection with comfort (Giusti Gestri, 2019). These informal adaptations reveal a gap between official compliance and lived user experience. Many riders expressed frustration at the lack of innovation and choice in the market, while medical staff demonstrated deep familiarity with vest models and certification protocols but lacked tools that supported effective emergency handling. This divergence between regulatory intention and practical usability underscores a fundamental disconnect between standards compliance and user-centered design (Forlizzi & Battarbee, 2004; Norman, 1988).

Collectively, these findings point to the need for a design-led transformation in safety vest development and standardization. Ergonomic redesign should explicitly address body diversity and gender differences, while modular and intuitive features could improve emergency access and medical response. Integrating adaptive materials and sensor-based systems would enable real-time monitoring and enhance both protection and post-incident care (Giusti Gestri, 2020; Berni & Borgianni, 2021). Most importantly, inclusive design processes should actively involve both primary users (jockeys) and secondary users (medical professionals) in the co-creation and evaluation of protective equipment, ensuring that design values (such as comfort, usability, and care) are embedded from the outset (Van der Velden & Mörtberg, 2015).

In summary, the evidence demonstrates that current safety vests fail to meet the functional, ergonomic, and procedural needs of their intended users. By illuminating these gaps, this case study strengthens the argument for a human-centered, participatory approach to safety design. The following chapter builds upon these insights to explore how core design values can be systematically embedded into both product innovation and the evolution of safety standards, transforming regulation from a constraint into a catalyst for progress.

Chapter 5.
The Evolving Role of the User: Innovation, Safety, and Smart Design

The Australian safety vest case study reveals more than ergonomic flaws or regulatory gaps; it highlights a fundamental shift in the relationship between users, technologies, and design systems. Traditional models treated users as passive recipients of standardized protection. Today, users are increasingly recognized as active participants and co-creators in the evolution of safety technologies (Bannon, 2011; Sanders & Stappers, 2014). This change reflects a broader movement in design research and human–technology interaction: the transition from *user-centered* to *user-empowered* frameworks, where personal expertise, lived experience, and embodied knowledge drive innovation.

In protective equipment, this evolution intersects with advances in smart materials, wearable technologies, and data-driven design (Giusti Gestri, 2023b; Dunne, 2010). Devices no longer merely shield—they sense, communicate, and adapt. For designers, this convergence offers opportunities for personalization, responsiveness, and inclusivity, but also raises ethical and practical questions. *How can emerging technologies enhance safety without creating exclusion or overburdening users with monitoring systems? How should design and policy frameworks evolve to support innovation while maintaining social accountability?*

This chapter reframes the user as an essential agent of innovation, not merely the endpoint of design but a dynamic force within an ecosystem connecting materials, data, and standards. Drawing on participatory design, socio-technical systems theory, and responsible innovation frameworks (Von Hippel, 2005; Stilgoe et al., 2013; Forlizzi & Battarbee, 2004), it situates jockey safety vests within a broader dialogue on the future of protective technologies. Viewed through this lens, safety design becomes both a technical and cultural practice, balancing regulation with creativity, care, and human insight.

5.1 Introduction

Protective technologies are no longer just about compliance, they are about collaboration, adaptation, and responsiveness. The evolution of jockey safety vests exemplifies this shift: users are active contributors whose experiences, insights, and tacit knowledge shape innovation. This perspective places the user at the heart of a network linking materials, data, regulatory frameworks, and design practices.

This chapter explores how users' lived experiences inform the development of next-generation safety technologies. By examining the interplay between human expertise and technological capabilities, we uncover opportunities for smarter, more inclusive, and adaptive designs. This approach challenges conventional boundaries in protective equipment, encouraging designers and policymakers to consider not only safety outcomes but also the ethical, cultural, and social dimensions of innovation.

5.1.1 Design in Protective Technologies

Insights from the Australian safety vest study extend beyond empirical observation, they illuminate the interdependence between human behavior, ergonomic needs, and material performance. Jockeys' feedback demonstrates that effective safety design emerges from ongoing dialogue between user and designer.

By centering users in the innovation process, designers can anticipate challenges, tailor functionality, and enhance inclusivity. Smart materials, sensor integration, and adaptive wearables transform protective gear from static shields into interactive, context-aware systems. While these advances offer personalization and adaptability, they also underscore the designer's responsibility to address ethical concerns, including privacy, accessibility, and equity.

5.2 User Participation as a Driver of Innovation

Traditional design approaches often treated user feedback as a final validation step. Contemporary frameworks, by contrast, position users as co-creators from the outset. Participatory design methodologies foster continuous interaction between designers and users, producing innovations grounded in real-world experience.

In jockey safety vests, this approach informs everything from fit and comfort to the integration of sensors and smart materials that anticipate risk. User involvement ensures that technological advances are meaningful, usable, and ethically responsible, guiding design through iterative cycles that reflect lived experience.

5.3 Navigating Ethical and Practical Challenges

While smart technologies expand possibilities, they also introduce new challenges. Data-driven systems can enhance safety but may also raise privacy concerns, exacerbate inequalities, or impose cognitive burdens. Addressing these challenges requires balancing regulatory compliance, technical innovation, and ethical foresight.

Designers must create tools that empower rather than control, and policymakers must adapt frameworks to support innovation while safeguarding equity, transparency, and human agency. Ethical foresight becomes as crucial as technical performance in shaping the protective technologies of tomorrow.

5.4 Towards a Smart, Inclusive, and Adaptive Future

The trajectory of protective technologies points to a future where devices are functional, intelligent, responsive, and inclusive. Centering user insight in design ensures that innovation balances technical performance, human experience, and social responsibility.

Jockey safety vests illustrate that true innovation is not only about materials or sensors but also about collaboration, empathy, and ethical reflection. By embracing participatory, data-informed, and socially aware design, protective technologies can evolve into tools that are both effective and empowering, offering a model for the future of human-centered innovation in safety systems.

As the chapter closes, it becomes clear that safety innovation does not occur in isolation. The relationship between design, regulation, and user agency shapes both the possibilities and the limits of protection. Understanding this interplay sets the stage for the next discussion: the paradox of protection, where standards, innovation, and ethical responsibility converge.

Chapter 6.
Innovation, Standards, and the Paradox of Protection

Safety standards are established to protect, ensuring that products meet minimum safety thresholds and providing a baseline for quality and performance. However, in the case of jockeys' safety vests, these standards (ARB 1.1998 and EN 13158) have inadvertently become barriers to innovation. While they aim to safeguard users, their rigid and prescriptive nature has stifled design evolution, excluded user diversity, and discouraged the adoption of new technologies.

A safety standard should not be a wall that innovation crashes into — it should be the floor that new ideas stand on.

Manufacturers of safety vests are compelled to design products that pass specific impact tests, coverage templates, and material specifications. While this ensures a degree of uniformity and reliability, it also creates a compliance trap. Designers prioritize test performance over user experience, limiting innovation to what can be certified rather than what is needed. Consequently, products become optimized for laboratory conditions, not real-world use, resulting in vests that are technically compliant but experientially deficient, such as uncomfortable, inflexible, and poorly suited to the needs of diverse users.

Product innovation in high-risk sports PPE such as thoroughbred horse racing is not merely a matter of technological advancement: it is a matter of safety, usability, and survival. This chapter critically examines the role of users in shaping innovation, with a particular focus on jockeys' safety vests. These vests occupy a complex space at the intersection of regulation, tradition, and emerging technology. In Australia, the regulatory framework (anchored in ARB 1.1998 and EN 13158) prescribes specific impact tests, coverage templates, and conditioning protocols that collectively constrain the shape, materials, and construction techniques of approved vests.

While these standards are intended to protect riders, their rigid application has fostered a culture of conservatism that resists ergonomic innovation and the integration of advanced materials. As discussed in Chapter 3, the Japanese Descente vest exemplifies this tension: despite its widespread acclaim among jockeys for comfort and flexibility, it was disqualified under Australian standards due to non-conforming design elements (Foote et al., 2014). This example highlights how regulatory systems can inadvertently suppress user-informed innovation, and how standards might evolve to better reflect the lived realities of athletes.

Current Australian safety vest standards also fall short in addressing emerging priorities in protective equipment design. Gender-specific ergonomics are notably absent, resulting in poor fit and restricted mobility for female jockeys and individuals with smaller body frames. Innovations such as smart materials and embedded sensors, which could enable real-time impact monitoring, adaptive protection, and improved medical response, remain excluded from existing protocols. Similarly, modular or detachable components, which could facilitate emergency access or allow for personalized adjustments, are not accommodated within current testing frameworks. These omissions reflect a broader disconnect between evolving technological capabilities and the static nature of regulatory standards. Even environmental adaptability, such as resilience to heat, humidity, or extended wear, is inadequately addressed. Overall, these gaps reflect a limited view of safety, focusing on measurable metrics rather than the nuanced, experiential realities of riding and emergency situations.

On a global scale, there are three primary international standards for equestrian body protectors: EN 13158, which is the European standard; and ASTM F1937 and ASTM F2681, which are the standards used in the United States. EN 13158 applies to body protectors for horseback riding in Europe, while ASTM F1937 and F2681 cover protective bodywear and headgear, respectively, for horse sports in the U.S. In some cases, standards developed for motorcycle protective gear are applied to equestrian equipment. These include EN 1621-2, EN 1621-3, and EN 1621-4. For instance, the Shadow back protector meets the EN 1621-2:2014 standard. It is important to note that while a back protector shields the spine, it does not offer rib protection like a full equestrian body protector. Notably, products bearing the CE mark have been assessed and meet the requirements of the Personal Protective Equipment (PPE) European Regulation 2016/425, confirming their compliance with European safety standards.

Three test regimes dominate the discussion:

• EN 13158 (Body protectors / Level 1 for jockeys): requires two impact configurations — a flat circular impactor (80 mm diameter) at ~25 J and a narrow-bar impactor (80 × 20 mm) at ~20 J — with pass/fail limits such as an average transmitted force not exceeding ~4 kN (derived historically from cadaver chest tests). EN conditioning uses moderate temperature/humidity cycles.

• SATRA (UK-derived test protocols): historically used a similar flat-impact test but with differing anvil geometry and multiple energy levels (including some higher-energy drops). SATRA ran whole-garment tests and additional weak-point tests and was influential in the original ARB formulation.

• ARB 1.1998 (Australia): originally based on SATRA but adapted for local climate. Crucially ARB requires elevated-temperature conditioning (40°C) and (unlike EN/SATRA) in practice has not retained the narrow-bar bending test as an explicit mandatory requirement in the same form; ARB also allows meeting only one of certain SATRA tests. That difference has important consequences for bending stiffness and segmented-panel designs.

Safety standards align on energy-absorption metrics, coverage templates, and pass/fail thresholds, but small differences (such as temperature conditioning and required test configurations) significantly influence which constructions and materials pass approval. While the Australian ARB 1.1998 standard established an early benchmark for protective vests, international frameworks such as EN 13158 have since evolved to reflect broader user demographics, improved ergonomics, and design innovation. Table 2 compares their key characteristics and illustrates how differing regional priorities influence product evolution.

Table 2. Comparison of Major Body Protector Standards: Australia, Europe, USA, and UK

Feature	**ARB 1.1998 (Australia)**	**EN 13158 (Europe)**	**ASTM F1937 (USA)**	**SATRA M38 (UK)**
Region of Application	Australia & New Zealand	Europe	United States	United Kingdom
Certified By	Standards Australia	European Committee for Standardization (CEN)	ASTM International	SATRA Technology Centre
Test Methods	Flat and rail impact tests; torso coverage; attachment security	Multiple impact tests (flat and bar); coverage; sizing; adjustability; ergonomics; material safety	Impact and coverage testing; attachment security; ergonomic evaluation	Impact and coverage testing specific to air vests; inflation reliability
Performance Levels	Level 1 only	Levels 1, 2, and 3 (*Level 3 = highest protection*)	Levels 1 and 2	Not formally specified
User Demographics	Professional jockeys; male and female riders	Riders of all ages and genders; includes youth sizes	Adult riders; professional and recreational users	Adult riders; primarily used with air vests
Climate Considerations	Considers Australian	No specific climatic testing	No specific climatic testing	No specific climatic testing

	climate conditions			
Regulatory Compliance	*Mandatory* for jockeys in Australia and NZ; enforced by racing authorities	Widely adopted; *voluntary* compliance across Europe and Australia	Widely adopted; compliance may vary by state	Common in UK; often used with EN 13158
Innovation Constraints	No revisions since 1998; rigid padding rules; lacks gender-specific ergonomics; excludes smart tech	Comprehensive requirements; limits detachable parts; strict sizing and coverage	Focus on minimum protection; limited adaptability for smart materials	Designed for air vests; limited standalone recognition
Allowances / Opportunities	Permits lightweight construction if compliant; simple flat-impact testing	Multiple protection levels; optional high-temperature testing; ergonomic focus	Accepts new materials; includes environmental conditioning tests	Encourages innovation in inflatable protection systems
Key Insight	Outdated but regionally enforced	Most comprehensive and up-to-date framework	Flexible but less prescriptive	Specialized; complements European standard

Australian safety vests for jockeys are shaped by regional differences, leading to a variety of standards across the globe. In Australia (and New Zealand), ARB 1.1998 was developed to account for local climatic conditions, while EN 13158, originating in Europe, has become the dominant standard across European countries. In the United States, ASTM F1937 is commonly applied, and in the United Kingdom, SATRA M38 is widely used. These variations reflect both environmental considerations and regional regulatory preferences.

When it comes to protection levels, the standards differ in their approach. EN 13158 defines three distinct performance levels (e.g., Level 1, Level 2, and Level 3) providing riders with options for varying degrees of protection. By contrast, ARB 1.1998 specifies a single performance level, Level 1, which may limit choices for riders seeking higher levels of safety. Design flexibility is another differentiating factor. ARB 1.1998 offers a relatively narrow range of design templates, which can constrain innovation and the development of new protective technologies. In comparison, EN

13158, ASTM F1937, and SATRA M38 allow for more diverse designs, accommodating evolving approaches to safety and rider comfort.

6.1 Why Australian standards (and their application) constrain innovation

There are four interacting mechanisms that freeze design choices:

1) *Prescriptive coverage templates and dimensional conformity.* EN/SATRA/ARB set coverage maps and maximum gap allowances between panels. Manufacturers design to the template rather than rethinking protection architecture; non-conforming silhouettes face uphill approval tests. The lack of an official, professional definition of "jockey safety vest" deepens this conservatism (Giusti Gestri, 2019).
2) *Approval-list market gatekeeping.* Racing Australia's "Approved Level 1" list restricts Raceday wear to a short catalogue of vests. Once a product family is approved the market becomes path-dependent; newer concepts (e.g., inflatable elements, novel composites, sensor-integrated textiles) must still pass tests that were conceived for foam panels. That discourages R&D investment by smaller innovators (Giusti Gestri, 2019).
3) *Test focus misaligned with real-world biomechanics.* Standard drop tests use rigid anvils and single-axis drop energy to measure transmitted forces (and the 4 kN threshold references old cadaver chest stiffness work). Real jockey falls involve multi-directional loading, rotational accelerations, crushing and trampling. Existing tests therefore measure a limited slice of safety performance and can miss failure modes (e.g., multi-axial spinal loading) that matter most in the field. Some Australian publications (Foote et al., 2011, 2014) specifically identify this representativeness gap.
4) *Conservative material assumptions baked into tests.* Tests were developed when foam strips covered with mesh were the dominant technology. Newer viscoelastic foams, high-strain composites, air-inflation systems, or smart fabrics behave very differently under the standard impacts and under conditioning, yet there is no common, accepted pathway to validate these materials within the regulatory frame. In the past decades, the Australian standards have seen minimal meaningful revision.

Together, these factors produce a regulatory environment that rewards small incremental changes (thickness, foam density) but penalizes radical alternatives (hexagonal mosaic panels, breathable composites, integrated sensors), even when such alternatives promise improved ergonomics and lower long-term injury risk.

6.2 Ergonomics, gender and missing user-centered design

Jockeys and co-dependent users have never been involved in the standards development and/or revisions. The result is a one-template-fits-all approach that neglects jockey posture ("Martini glass" crouch), female body shapes, emergency removal needs, and long-duration thermal comfort. Multiple jockeys participating in the Australian case study have reported restricted movements, back/neck exposure, and difficulty in emergency extraction, problems that the D-UC method and co-design could address. A more recent publication (Giusti Gestri, 2023b) reinforces this finding and explicitly proposes integrating wearable sensors as a route to better protection and post-incident triage. Particularly, both jockeys and medical staff are receptive to embed sensors, advanced materials, and explore prototypes for:

- Objective crash-detection and impact magnitude recording,
- Reduce the recovery time,
- Decrease the medical and/or insurance costs,
- Real-time physiological telemetry (e.g., heart-rate, orientation), and
- Data-driven injury registries to support standard revision and targeted R&D

However, standards currently lack a certification pathway for sensor-bearing vests. Questions regulators will raise include device robustness, data security, failure-safe behavior, and whether sensors change the passive protective performance (e.g., compromise foam continuity). Without a dual-pathway approach (a conservative "legacy" track and an innovation "sandbox"), manufacturers and developers are stuck: their innovations can't be trialed in races at scale because the garment as a whole must still satisfy legacy mechanical pass/fail tests. Hence, Australian standards are blocking innovation in the jockeys' safety vests. A design-values paper explicitly recommends design-guides and co-design processes to bridge this gap (Giusti Gestri, 2024).

6.3 Brief historical timeline (key inflection points)

- 1996: SATRA tests and UK/BETA approaches settle into practice (influential for later standards).

- 1998: ARB 1.1998 introduced in Australia (based on SATRA, adapted for heat/wet conditions). Vests became compulsory in Australia.
- 2009: EN 13158 consolidation (BETA adoption/re-implementation in UK).

- 2014: ARB ceases SATRA use after review; operative standards remain ARB 1.1998 and EN 13158. Foote et al. highlight omissions and recommend changes (temperature conditioning, spine coverage).

- 2019 - 2025: same Australian standards are in use, and many calls for improvements have been made, but nothing has changed yet (Giusti Gestri, 2023 a, b, 2024; Dorries, 2025).

6.4 Practical recommendations (for a standards revision / innovation pathway)

Reforming the standards that govern jockey safety vests demands more than another round of technical amendments. It requires a cultural and structural shift in how innovation is tested, approved and absorbed into the rules of racing. The first and perhaps most transformative step is the creation of a dual-track approval pathway. Under this model, a *legacy track* would continue to certify traditional foam-based vests using the familiar impact tests and coverage templates. Alongside it, an *innovation sandbox* would permit new materials, segmented architectures and sensor-integrated systems to be evaluated under a combination of controlled laboratory and monitored field conditions. Data gathered in this sandbox (e.g., quantitative impact readings, physiological responses and user feedback) would flow into a scheduled review process, ensuring that real-world evidence actively shapes the next revision of the standard. This idea, proposed in *Assessment of Design Values in Designing Product Standards* (2024), reframes the standard as a living design tool rather than a fixed rulebook.

A second priority, strongly argued in a doctoral thesis (Giusti Gestri, 2019), is to expand the representativeness of impact testing. The current single-axis drop tests, inherited from earlier SATRA and EN protocols, capture only a fraction of the forces experienced when a jockey falls. Future test programs should employ finite-element modelling and multi-axis drop or rotation rigs to reproduce torsional, shear and crush effects more faithfully. Complementary procedures, thermal conditioning, repeated-load ageing and trampling simulations, would account for the ways heat, sweat and repeated impacts degrade materials over time. Data from instrumented field trials, using the smart-vest sensors now being prototyped, could then be used to calibrate these laboratory models, closing the gap between mechanical testing and real kinematics.

Standards must also begin to treat ergonomics and co-dependent user needs as core safety criteria. Notably, the fit and posture of the vest have direct implications for both performance and injury management. Mandatory posture-specific fitting trials should replicate the jockey's crouched riding position, while removal tests should be co-designed with race-day medical staff to guarantee swift access after a fall. Comfort and accessibility are not secondary features; they are part of protection itself. To reflect the diversity of today's riding cohort, future standards should also introduce gender-specific templates or modular sizing that accommodate different torso shapes rather than forcing all riders into a single silhouette.

Equally important is the introduction of a data-driven validation requirement for innovation. As previously suggested by the author (Giusti Gestri, 2023b), any new protective concept should be supported by biomechanical laboratory evidence and at least one monitored track trial, ideally conducted during non-competitive sessions. Manufacturers would then contribute anonymized crash and performance data to a central research registry. Over time this evidence base would reveal which design features correlate with reduced injury severity, allowing standards committees to recalibrate protection thresholds on the basis of empirical data rather than expert intuition alone.

Finally, meaningful reform depends on governance that is inclusive, transparent and cyclically refreshed. Following the example of the British Equestrian Trade Association, standards committees should integrate the perspectives of active riders, thus, jockeys, medical practitioners and materials scientists alongside regulators and manufacturers. Their combined experience would ensure that laboratory metrics remain aligned with on-track realities. Each revision cycle (ideally every three to five years) should conclude with a public report summarizing the evidence that prompted each change, creating accountability and trust within the industry.

Together these actions chart a practical route toward standards that are adaptive, evidence-based and humane. They replace a compliance culture with a learning culture, one that values data, design insight and lived experience equally. By embedding innovation within the framework rather than outside it, Australian racing can move from a reactive safety posture to a genuinely progressive model in which every new vest, and every new dataset it generates, helps shape the next generation of protection. Notably, the cost of this stagnation is substantial. Riders are compelled to use vests that compromise comfort and performance, medical staff encounter delays during critical interventions, and manufacturers are disincentivized from investing in research and development. Furthermore, when users perceive equipment as cumbersome or untrustworthy, compliance becomes superficial, undermining the protective intent of the standards themselves.

To overcome these constraints, a design-led approach to standards is required. Rather than prescribing specific design inputs, regulations should define desired safety outcomes while allowing multiple pathways to compliance. This approach encourages iterative user testing, feedback integration, and pilot programs for new materials and technologies. Crucially, it promotes inclusive co-design, involving designers, riders, and medical professionals in shaping both the product and the standards that govern it. By framing safety as contextual, experiential, and dynamic, design-led standards create the conditions for equipment that not only protects but also enhances usability, inclusivity, and trust.

6.5 Summary

Current Australian jockey-vest standards were never meant to be static, yet decades of incremental revision have made them so. Reform now requires an approach that blends evidence, empathy and experimentation. A dual-track approval system would let innovation flourish without jeopardizing safety; expanded multi-axis testing would mirror real-world falls; ergonomics and gender-specific fit would return the rider's body to the center of design; and data-driven trials would ground every rule in lived performance.

With broader governance and scheduled, transparent review, standards could finally evolve at the same pace as the technologies they regulate. At its heart, this shift is cultural. It asks regulators to see standards not as barriers but as *platforms for discovery*: frameworks that learn, adapt and improve through the very data they generate. In doing so, the system that once stifled innovation could become the engine that sustains it.

In conclusion, the paradox of protection lies in the tension between compliance and lived experience. Current standards, while ensuring measurable safety, inadvertently suppress innovation, limit user diversity, and constrain technological advancement. Moving forward, the challenge is to create regulatory frameworks that balance rigorous protection with flexibility, adaptability, and human-centered design.

Chapter 7.
From Compliance to Creativity – Conceptual Proposals for Next-Generation Safety Vests

The discussion so far has shown that safety in horse racing is shaped as much by bureaucracy as by biomechanics. Previous chapters revealed how regulatory frameworks, testing regimes, and entrenched design traditions combine to produce a culture of *minimum compliance*: a system where manufacturers design to pass a test rather than to protect a body. Jockeys and medical personnel described equipment that meets every formal criterion yet still feels uncomfortable, restricts movement, and complicates emergency response. The result is a paradox familiar to many safety-critical industries, devices that are technically safe but experientially inadequate.

The persistence of this paradox invites a shift in perspective. Instead of viewing standards as the fixed endpoint of design, they can be understood as one variable in a broader ecosystem of practice, an ecosystem that includes users' bodies, their working environments, and the emotional dimensions of trust and comfort. The challenge, then, is not merely to tighten tests or adjust templates, but to re-imagine what *protection* means in an era of digital materials, wearable sensors, and participatory design.

This chapter takes up that challenge. It asks how creative methods, drawn from design research and human-centered innovation, might reconfigure the relationship between the rider and the rules that govern their safety. By translating empirical insights into conceptual proposals, the discussion moves from diagnosis to design, illustrating how a vest might evolve from a static object of regulation into a dynamic interface for wellbeing, performance, and identity.

7.1 Introduction: Rethinking Protection

Throughout the preceding chapters, a recurring tension has surfaced between the restrictive power of standards and the lived realities of the people who depend on them. Compliance delivers measurable safety, yet it often narrows the scope for imagination, usability, and inclusion. If the existing standards operate as a ceiling that limits innovation, then design must become the means of lifting that ceiling.

Building on the empirical findings discussed in the previous chapters 4 to 6, this one proposes conceptual directions for a new generation of jockeys' safety vests, designs that place *users* rather than *rules* at their center. These proposals draw on principles from D-UX, which frames every artefact as part of a relational network connecting the user, the technology, and its context (Sanders & Stappers 2014; Norman 2013). In this framework the vest ceases to be merely a mandatory shield: it becomes a responsive, empowering interface that enhances safety, comfort, and confidence simultaneously. Through this design-led lens, the chapter explores how ergonomics, modularity, and intelligence can coexist with rigorous protective performance to bridge the long-standing divide between compliance and creativity.

7.2 Design Goals for the Next-Generation Vest

The design agenda that follows arises directly from the deficits identified earlier: poor fit, limited adaptability, and inadequate interaction between riders and medical personnel. The first priority is *ergonomic fit*. Vests must accommodate diverse body shapes and riding postures, using gender-specific contours and adjustable padding zones. Research has shown that standardized sizing disadvantages female and smaller-framed jockeys, often forcing them to compromise either comfort or protection (Brisbine et al., 2022; Giusti Gestri, 2019).

A second goal concerns *modularity*. Rather than treating the vest as a sealed object, it should function as an adaptable system, one that can be repaired, customized, or even partially removed under emergency conditions. Modularity also enables maintenance and sustainability, reducing the waste associated with short replacement cycles.

Third, the vest should evolve into a *smart* object. Embedded sensors capable of recording impact magnitude, heart rate, and body temperature could supply live data to trackside medical teams, providing an immediate picture of injury risk (Kim, 2025; Giusti Gestri, 2023b). These same datasets would, in the longer term, support evidence-based revisions to testing protocols. Instead, the fourth ambition is to achieve *lightweight protection*. Advances in viscoelastic foams, mesh composites, and phase-change textiles now make it possible to reduce bulk while maintaining, or even improving, energy attenuation (Peng et al., 2024; Tang, 2025; Tamjid et al., 2025). Comfort is not a luxury; it is a precondition for consistent compliance.

Equally significant is aesthetic appeal. Allowing riders to choose colors, patterns, and finishes fosters personal identification with the equipment; emotional engagement with an artifact (through its appearance and feel) strengthens habitual use. Research in wearable technology shows that aesthetics and perceived comfort influence long-term use, and studies in PPE noncompliance suggest that design factors (including look and comfort) matter for consistent usage (Aryana et al., 2024; Nagare, 2025). Finally, *medical accessibility* must be designed in from the outset. Intuitive tear-away sections and clearly coded release points enable medical staff to remove the vest swiftly without damaging its structural layers, a response to repeated failures observed in incident analyses (Kim, 2025). Together, these intertwined goals redefine protection as an active relationship between body, technology, and context rather than a static set of physical barriers.

7.3 Conceptual Features and Prototyping Directions

Realizing these goals demands more than theoretical intent. Several conceptual features illustrate how they can materialize in practice. Magnetized closures could replace noisy Velcro and stiff zippers, providing both rapid emergency release and a cleaner ergonomic profile. Modular panels attached through these magnetic interfaces would allow riders to alter coverage density according to race type or climate conditions, echoing Norman's (2013) emphasis on adaptability as a driver of usability.

Within the vest's layered structure, flexible biometric sensors could record physiological and impact data in real time. These lightweight components would transmit information wirelessly to medical staff, forming a direct feedback loop between athlete and caregiver. Design templates based on updated anthropometric datasets would ensure that male and female body geometries are addressed equally, while flexible spine extensions would protect the vulnerable "Martini-glass" posture identified in previous analyses. Complementing these physical refinements, breathable smart fabrics employing phase-change or moisture-responsive fibers could regulate micro-climate temperature, maintaining comfort in the demanding heat of Australian.

Visual identity remains integral: subtle color accents, embossed textures, or personalized motifs can transform an obligatory uniform into a mark of professional pride. Finally, coded tear-away zones (visually marked yet structurally discrete) would guide medics to access points without compromising integrity during normal use. Together, these ideas outline a vest that communicates as much as it protects: intelligent, expressive, and humane.

7.4 Design Scenarios and Early Explorations

Doctoral research undertaken for this project generated a suite of exploratory sketches and mock-ups developed through workshops with jockeys and medical professionals. Concepts such as hexagonal foam tessellations demonstrated how flexibility and impact distribution could coexist. Spine-protecting tail extensions were modelled to maintain posture support, while modular shoulder assemblies offered repairability and adjustable padding. Integrated data ports hinted at how sensor modules might connect to real-time monitoring systems.

Feedback from focus groups revealed strong enthusiasm for the potential of these prototypes, particularly among riders who had experienced discomfort or restricted movement in conventional vests. Medical personnel emphasized the value of predictable release mechanisms and clearer access paths. These dialogues validated the central premise of participatory design: that meaningful innovation arises when users and experts share authorship of solutions (Sanders & Stappers 2014).

7.5 From Concept to Practice: Testing and Collaboration

Transitioning these concepts into functional prototypes requires an iterative and collaborative process. The first stage involves co-design workshops where riders, medics and designers jointly refine ergonomics and usability. Rapid prototyping using additive manufacturing and textile fabrication follows, allowing material testing and real-time adjustment. Field evaluations under simulated race conditions then measure comfort, mobility, and emergency accessibility, while structured feedback loops ensure each prototype evolves from lived experience rather than assumption.

Crucially, collaboration with standards bodies must begin early. Flexible certification pilots could focus on outcome-based metrics (e.g., impact attenuation, accessibility, data reliability) rather than prescriptive templates. Such an approach aligns with Norman's (2013) argument for human-centered design systems and with emerging evidence from wearable-technology research demonstrating that innovation and safety can coexist when regulation measures performance instead of form.

7.6 Summary and Transition

This chapter has reframed the safety vest as a site of creative possibility rather than regulatory constraint. By uniting ergonomic refinement, modular construction, intelligent sensing and aesthetic expression, next-generation designs promise to transform protection from a passive obligation into an active, empowering experience. They illustrate how compliance and creativity need not be opposing forces but complementary components of a more mature design culture, one that measures success not only by impact scores but also by comfort, inclusion and user trust.

The next chapter extends this vision beyond the racetrack, considering how these design values can inform protective equipment in other high-risk sports and occupational environments. Through that broader lens, the principles developed here, co-design, adaptability, and evidence-based evolution, emerge as universal strategies for aligning human experience with technical protection.

Chapter 8.
Beyond Horse Racing: Lessons for High-Risk PPE

While horse racing provides a highly visible illustration of high-risk PPE, analogous challenges appear in many other domains. Consider firefighters navigating smoky, unstable structures: their gear must shield them from heat and impact while allowing agile movement and rapid decision-making. Surgeons and laboratory technicians face similar trade-offs, balancing protective barriers against the need for dexterity, visibility, and precision. Even industrial workers on construction sites encounter the same tension between standardized equipment and individual comfort, with poorly fitted or cumbersome PPE sometimes creating more danger than it prevents.

Across these fields, one recurring pattern emerges: innovation often arises not from regulation alone, but from iterative, UCD that prioritizes real-world experience. Modular components, adjustable sizing, and task-specific customization (concepts lacking in the equestrian world, especially in the thoroughbred horse racing field) have proven transformative when applied elsewhere. By examining these parallels, we begin to see that the horse racing vest is not merely a niche artifact; it embodies principles that can inform a systematic rethink of protective gear across contexts where risk is high and human performance is critical.

This perspective also highlights a subtle but powerful insight: effective PPE cannot be imposed solely from the top down. Regulatory frameworks, while essential, must coexist with mechanisms for feedback, experimentation, and adaptation. Only by cultivating a culture that values design thinking and iterative improvement can high-risk industries evolve their protective equipment from a compliance checkbox into a genuinely enabling tool for safety and performance.

8.1 Introduction: From Niche to Norm

Although this book takes the jockey's safety vest as its central case study, the arguments developed throughout reach far beyond the racetrack. The design values, ergonomic priorities, and innovation barriers discussed earlier echo across a wide landscape of PPE. Wherever human beings work or compete in high-risk environments (on building sites, in hospitals, in mines, or on ski slopes) the same fundamental tensions recur safety versus mobility, standardization versus individual fit, and regulation versus responsiveness.

The racing vest therefore serves as a lens through which to view a broader dilemma: how can societies protect their workers and athletes without stifling the creativity that could make protection more effective? The insights gained from reimagining equestrian protection (dual-track standards, modular design, user-centered innovation, and dependency-based usability) offer a transferable vocabulary for change. This chapter expands that conversation to examine how design-led thinking can reshape PPE across multiple sectors, arguing that innovation in protection must evolve from a niche experiment into a cultural norm.

8.2 Cross-Sector Challenges in Personal Protective Equipment

Across the spectrum of high-risk activity, protective equipment exists in a delicate balance between *control* and *freedom*. In motorcycling, helmets and jackets must absorb high-energy impacts without limiting neck rotation or airflow. In skiing and snowboarding, lightweight protectors must combine warmth, breathability, and freedom of movement in sub-zero conditions. Cyclists face the added demand of aerodynamics and fashion appeal: equipment must safeguard yet remain sleek, stylish, and compatible with performance tracking devices.

Beyond sport, industrial workers navigate even harsher contradictions. Construction and mining personnel wear protective gear for long durations under heat, dust, and vibration. Comfort, flexibility, and gender inclusivity become as critical as resistance to puncture or flame. In healthcare, PPE represents a different but equally complex negotiation. Masks, gowns, and gloves are essential for infection control, yet they can impede communication, tactile sensitivity, and empathy: qualities that define good clinical practice. These examples underline that the challenges identified in equestrian safety (such as the dominance of outdated standards, poor fit, limited innovation, and exclusionary sizing) are systemic rather than sector-specific (Al-Bayati et al., 2023; Aryana et al., 2024; Malta et al., 2024).

In each domain, standards tend to measure material endurance and mechanical strength more readily than they assess usability, psychological comfort, or social acceptance. As Sanders and Stappers (2014) note, the absence of user participation in standardization leads to artefacts that protect bodies while neglecting experiences. There are similarities with healthcare PPE, where compliance stays high, but enthusiasm for adoption stays low, indicating that functional adequacy does not equal human usability. This cross-sector evidence points to a critical insight: the problem is not with the concept of standards but with their temporal inertia. In fast-evolving technological and social landscapes, rules built on the materials and anthropometry of the past quickly become barriers to progress. To move forward, industries must embrace standards as *living instruments* that co-evolve with design research, user feedback, and cultural change.

8.3 Modularity, Smart Integration and Participatory Design

The re-imagined jockey vest provides a template for how modularity, intelligence, and participation might operate together in other fields. Modularity allows protective systems to adapt to diverse tasks and bodies: panels can be removed, replaced, or upgraded without discarding the entire garment. This approach already appears in military body armor and firefighter ensembles, where detachable components accommodate seasonal or situational variation. Extending such modular thinking to industrial or healthcare PPE could dramatically improve both sustainability and user comfort.

Smart integration introduces another dimension. The embedding of sensors, whether for impact detection, temperature monitoring, or physiological feedback, has moved rapidly from laboratory prototypes to consumer devices. Helmets that register concussion forces, gloves that track hand strain, or masks that log air quality all illustrate the potential of data-driven safety systems (Vasile et al. 2023). Yet these technologies also expose a regulatory gap: most PPE standards have no mechanism for validating digital performance. The lesson from the jockey-vest research is that approval processes must evolve to assess the *function* and *resilience* of smart systems rather than reject them for not fitting traditional templates.

Participation completes the triad. The D-UX framework introduced in earlier chapters is particularly valuable in contexts such as emergency response or surgery, where multiple users rely on the same equipment in sequence. A respirator, for instance, affects not only the wearer but also colleagues who need to hear and understand them; a medical gown influences not only the clinician's movement but also the patient's sense of care and dignity. Applying D-UX means mapping these interdependencies, designing for collective experience rather than isolated individuals (Norman 2013; Sanders & Stappers 2014).

Participatory design methods (e.g., co-creation workshops, scenario simulations, and iterative prototyping) should therefore become central to PPE development. Involving firefighters in testing new heat-resistant textiles or nurses in redesigning surgical masks can yield immediate, practice-based improvements. The participatory process also generates ownership: when users help shape their gear, they are more likely to trust and maintain it.

8.4 Policy, Procurement, and Cultural Shifts

Design-led innovation, however, cannot thrive in isolation from institutional structures. The most elegantly conceived prototype will falter if procurement systems, funding models, and regulatory cultures remain conservative. Many public and private organizations still equate safety with compliance rather than with performance. Procurement criteria often privilege cost and certification over user comfort, lifespan, or adaptability.

To address systemic shortcomings in PPE design and deployment, design expertise must be embedded early in policy formation and procurement processes. Governments and industry associations could initiate pilot programs where designers collaborate directly with end users to evaluate new PPE concepts prior to formal tendering (Aryana et al., 2024). Funding mechanisms should support iterative testing and post-market evaluation, ensuring that field-based insights inform both product development and standard-setting (Kundu et al., 2025). Incorporating explicit design-value metrics (such as ergonomics, accessibility, and emotional acceptance) into procurement guidelines would help shift evaluation criteria from purely technical specifications to human-centered measures (Rotinen, 2022). Institutional collaboration is equally vital. Cross-sector partnerships can accelerate knowledge transfer: innovations in sports textiles may enhance emergency services gear, while sensor systems developed for healthcare can inform construction safety protocols. Interdisciplinary design exchanges have been shown to increase resilience and reduce duplication of research efforts (Duryan et al., 2020; Shi et al., 2021; Schimpf et al., 2024).

Policy frameworks must therefore recognize design not as a cosmetic add-on, but as a strategic component of safety infrastructure (Giusti Gestri, 2024). Hence, cultural transformation is equally essential. Standards bodies and regulators must cultivate an ethos of curiosity rather than caution. Instead of viewing novel materials or digital systems as threats to reliability, they can treat them as opportunities for iterative improvement. Training programs for inspectors and procurement officers could include basic design-thinking modules, encouraging a mindset of evaluation through experimentation. Over time, such initiatives would produce a workforce capable of bridging technical and human factors, engineers who understand empathy, and designers who grasp regulation.

The payoff is significant. PPE that is comfortable, inclusive, and trusted is more likely to be worn correctly and consistently. The economic implications are equally positive: improved durability and user satisfaction reduce replacement cycles and injury costs. The shift from compliance to creativity thus becomes not only ethically compelling but also economically rational.

8.5 Case Illustrations: Translating lessons across domains

To make these ideas tangible, consider three scenarios that echo the lessons of the redesigned jockey vest.

1. Motorcycling and urban mobility. Modern motorcyclists navigate congested cities where accidents often involve rotational rather than linear impacts. Helmet standards, however, still rely mainly on vertical drop tests. By integrating multi-axis testing and smart sensors—principles borrowed from equestrian research—helmet manufacturers could measure and mitigate real-world crash dynamics more accurately. Modular, magnetically fastened padding systems could also address the persistent issue of poor ventilation without compromising structural integrity.

2. Healthcare PPE and pandemic preparedness. The COVID-19 crisis exposed profound weaknesses in PPE design: fogging visors, bruising straps, and gowns that tore under stress. Applying D-UX principles would treat healthcare teams as interdependent users whose communication, comfort, and empathy are part of safety itself. Smart fabrics that monitor humidity and temperature could prevent heat exhaustion during long shifts, while modular attachment systems would simplify donning and doffing procedures—reducing contamination risk and fatigue.

3. Construction and mining. Workers in these sectors often modify their gear unofficially (e.g., cutting sleeves, removing padding) to cope with discomfort. A modular approach could formalize this adaptability. Components could be swapped according to weather or task, and embedded accelerometers could record impact and vibration exposure over time. Data collected would inform both occupational health policies and insurance frameworks, turning everyday wear into a continuous safety audit.

These brief examples demonstrate that the insights derived from horse racing (e.g., flexibility, smart feedback, and participatory design) are readily transferable. They illustrate how creative design, when aligned with human experience, can reconcile safety and agency across diverse environments.

8.6 Summary

The lessons of equestrian safety extend far beyond a single sport. They expose structural problems shared across all forms of high-risk protection: outdated standards, narrow definitions of safety, and limited dialogue between users and regulators. By applying design-led methods (such as, modularity, smart integration, participatory development, and inclusive policy reform) industries can transform PPE from a static compliance object into an adaptive system of care. Reconceptualizing protection in this way also reframes design itself. No longer confined to aesthetics or ergonomics, design becomes a strategic practice that links innovation, evidence, and empathy. The same approach that produced the next-generation jockey vest can inform the next generation of respirators, helmets, gloves, and gowns.

As we move into the concluding part of this book, the focus shifts from artefacts to institutions. The final chapters examine how to embed these design values within the governance of standards themselves, ensuring that safety frameworks evolve with human experience rather than lag behind it. Only when standards become dynamic, evidence-driven, and participatory will protection truly move *beyond compliance* to become a form of creative collaboration.

Chapter 9.
Rethinking Standards Through Design

Standards are more than technical specifications, they are the backbone of safety, shaping how products, practices, and behaviors are evaluated and regulated. Yet, despite their central role, they are often viewed as static, distant, and removed from the realities of users and designers on the ground. This chapter challenges that perception, proposing a shift in perspective: standards as dynamic, design-informed frameworks that evolve alongside technological innovation, user experience, and societal expectations. By rethinking standards in this way, we open a space for collaboration between regulators, designers, and users, transforming compliance into an active, creative, and context-sensitive process.

9.1 Introduction: Standards as Living Frameworks

Standards are often perceived as rigid, technical documents, fixed rules that define compliance. In reality, they are social constructs, shaped by institutional priorities, assumptions about users, and regulatory conventions. Traditional approaches treat standards as immutable, emphasizing prescriptive measurements over context, lived experience, or adaptability.

This chapter argues for reimagining standards as *living frameworks*: responsive systems that evolve alongside user needs, technological advances, and design innovation. By embedding human-centered design values into the standard-making process, safety systems can become inclusive, adaptable, and effective, promoting not only protection but also usability, comfort, and trust (Norman, 2013; Sanders & Stappers, 2014). Rather than being endpoints of compliance, standards can serve as catalysts for innovation, guiding design toward meaningful real-world outcomes.

9.2 Limitations of Static Standards

Current safety vest standards, such as ARB 1.1998 and EN 13158, exemplify the constraints of static regulation. They often rely on outdated assumptions about user demographics and body types, prescribe rigid templates that exclude modularity and emerging smart technologies, and limit meaningful stakeholder engagement, particularly from end users and co-users such as medical professionals. Testing protocols frequently prioritize lab-based, measurable performance, overlooking the lived experiences of jockeys in real-world conditions (Webster, 2010). In effect, safety becomes defined by quantifiable metrics rather than contextually grounded effectiveness. Standards are more than technical specifications: they are the backbone of safety, shaping how products, practices, and behaviors are evaluated and regulated. Yet, despite their central role, they are often viewed as static, distant, and removed from the realities of users and designers on the ground.

This chapter challenges that perception, proposing a shift in perspective: standards as dynamic, design-informed frameworks that evolve alongside technological innovation, user experience, and societal expectations. By rethinking standards in this way, we open a space for collaboration between regulators, designers, and users, transforming compliance into an active, creative, and context-sensitive process. These challenges are not unique to equestrian PPE. Across high-risk industries, protective equipment often favors standard compliance over comfort, mobility, and inclusivity, resulting in reduced adoption and, in some cases, diminished actual protection (BSI Group, 2025; Follett, 2025). Static standards, while providing certainty, risk creating systems that are safe on paper but misaligned with the realities of users' needs.

9.3 Toward Design-Informed Standards

Design-informed standards reframe safety as contextual, experiential, and dynamic, integrating human-centered principles into regulatory frameworks. Key strategies include:

- *Empathic research:* Conducting field studies, interviews, and participatory observations to understand user and co-user needs.
- *Iterative prototyping:* Testing usability alongside technical performance, allowing real-world feedback to guide refinement.
- *Participatory processes:* Engaging designers, users, regulators, and industry stakeholders in co-creating standards.
- *Flexible certification pathways*: Enabling multiple routes to compliance to accommodate innovation, modularity, and emerging technologies (Norman, 2013; Sanders & Stappers, 2014).

Through these strategies, standards shift from prescriptive, metric-driven rules to adaptive guidelines that recognize the diversity of users, environments, and operational conditions. Safety becomes a collaborative, evolving practice rather than a fixed requirement.

9.4 Embedding Design Values in Standards

Central to this transformation is the systematic integration of design values, like comfort, usability, inclusivity, and responsiveness, into both evaluation and development processes. Standards can achieve this by:

- Expanding evaluation criteria to include ergonomic fit, emergency accessibility, and user satisfaction, alongside technical performance.
- Moving documentation beyond prescriptive instructions, offering guidance on how products should *feel* and *function* in real-world conditions.
- Incorporating scenario-based testing to simulate realistic events, such as falls, medical interventions, and environmental stressors, ensuring standards reflect operational realities.
- Ensuring meaningful stakeholder representation, including users, co-users, and designers in the revision and approval process.

Embedding these values shifts the focus from mere compliance enforcement to collaborative safety design, aligning regulatory objectives with human experience and operational demands.

9.5 Institutional and Policy Implications

For design-led standards to succeed, systemic support from institutions and policymakers is essential. This support begins with the establishment of interdisciplinary working groups that bring together designers, engineers, regulators, and end users to guide the development of new standards. It also involves positioning design research as a core component of safety innovation, ensuring that creativity, empathy, and human-centered thinking inform technical and regulatory processes.

Pilot programs play a crucial role in this ecosystem, providing opportunities to test alternative compliance models and modular technologies in real-world contexts. At the same time, international collaboration should be encouraged to harmonize standards globally while preserving the flexibility needed to address local conditions and priorities. Equally important is the commitment to open access, sharing testing data and user feedback to support continuous refinement of standards in line with technological advances, societal expectations, and the diversity of user needs (Kampfmann et al., 2023; Sarkki et al., 2020). By embedding these practices within institutional frameworks, design-led standards can evolve as adaptable, forward-looking systems, dynamic tools for innovation rather than static sets of rules.

9.6 Conclusion: Standards as Catalysts for Innovation

Reconceptualizing standards as living, design-informed frameworks transform regulatory systems from rigid prescriptions into adaptive, human-centered tools. Safety becomes not just a technical requirement but a holistic system integrating comfort, trust, usability, and inclusivity.

This approach offers broader applications across sports, healthcare, and industrial domains, demonstrating that designing for safety is inseparable from designing for dignity, responsiveness, and meaningful adoption (Norman, 2013; Sanders & Stappers, 2014). By shifting the focus from compliance to collaboration, standards can guide innovation, ensuring that protective technologies evolve in step with both user needs and technological potential.

Looking ahead, the focus moves from the design of standards to the broader paradox of protection: the complex interplay between innovation, regulation, and user safety. Chapter 10 builds on these insights, exploring how design-led approaches can transform protective equipment, align safety with human experience, and guide both product development and policy frameworks in practice.

Chapter 10.
Conclusion: Designing for Safety, Designing for Change

As Chapter 9 illustrated, treating standards as living, design-informed frameworks transform them from rigid compliance tools into catalysts for innovation. Building on this insight, Chapter 10 reflects on how design-led approaches can reshape protective equipment, align safety with human experience, and guide both product and policy development.

10.1 Rethinking Safety and Standards

This book began with a simple insight: design feasibility does not reside in spreadsheets; it emerges in the parts you can test, move, and adapt. Using jockeys' safety vests as a lens, we explored how essential design values (like comfort, usability, inclusivity, and responsiveness) are frequently absent from product standards, despite being critical for real-world safety and adoption.

The Australian jockeys' safety vests case study highlighted a persistent problem: existing standards are not only outdated but also structurally resistant to innovation. Compliance often takes precedence over lived experience, rigidity over adaptability, and technical metrics over human needs. The consequences are tangible: discomfort, reduced performance, exclusion, and missed opportunities for meaningful innovation.

Reimagining standards as adaptive, human-centered frameworks offers a way forward. By integrating scenario-based testing, participatory stakeholder engagement, and user-focused documentation, safety systems can align protection with user experience. This approach ensures that standards reflect the realities of high-risk environments rather than idealized laboratory conditions, fostering both innovation and safety.

10.2 The Transformative Potential of Design-Led Approaches

Design-led strategies have the power to transform protective equipment from static compliance artifacts into adaptive, intelligent systems. The D-UX framework illustrates how design can address the needs of primary users, such as jockeys, while simultaneously considering co-dependent users, including medical professionals, whose interaction with equipment can be critical during emergencies. This dual perspective ensures that equipment supports not only individual performance but also broader operational safety.

Key principles such as modularity, smart integration, and iterative prototyping allow protective equipment to evolve alongside technological advances, accommodate individual variability, and facilitate rapid, context-sensitive responses. When these design values (such as, comfort, usability, inclusivity, and responsiveness) are embedded into both product development and standard-setting processes, they foster trust, adoption, and meaningful engagement among users.

Ultimately, design-led approaches redefine safety. It becomes more than technical compliance or protective functionality; it becomes experiential, relational, and empowering. Safety is understood as a dynamic interaction between people, technology, and environment, where standards and products evolve together to meet real-world challenges. By centering human experience alongside regulatory rigor, design-led strategies chart a pathway toward innovation that is not only safe but also equitable, adaptive, and forward-looking.

10.3 Designing for Change, Dignity, and Future Practice

The ultimate goal extends far beyond jockeys' vests: it is about designing for people, trust, and adaptability. Recognizing diverse bodies, roles, and operational contexts ensures that equipment supports performance while respecting dignity. Recent research expands the notion of who counts as a user in safety design. In equestrian sports, for example, the concept of co-dependent users highlights that safety outcomes depend not only on the wearer but also on those who interact with the equipment during critical moments (for instance, medical staff, stewards, or fellow riders). This broader perspective shifts design priorities from individual protection toward a system of shared safety, where every stakeholder plays a role in risk management and emergency response.

Empirical insights from the individual semi-structured interviews, focus group, and field observations reveal persistent gaps in existing safety equipment. Many jockeys reported discomfort, poor fit, and restricted movement, with female riders particularly affected by designs that overlook gender-specific ergonomics. Medical professionals noted challenges in accessing injured riders when certain protective gear hindered emergency interventions. These observations underscore how design choices directly influence not only comfort and performance but also the efficiency and effectiveness of post-incident care.

Current Australian standards and regulations, such as ARB 1.1998 and EN 13158, remain limited in scope, focusing largely on impact resistance and material compliance. They often overlook emerging needs (including body diversity, modularity, and the integration of smart materials or wearable technologies) that could transform protective design. By prioritizing conformity over adaptability, such standards inadvertently constrain innovation and slow the adoption of advanced, data-informed solutions.

Design-led strategies, exemplified by the D-UX framework, provide a structured framework for developing adaptive, user-centered, and technologically responsive protective equipment. Modularity, smart integration, and iterative prototyping allow products to evolve with technological advances, accommodate individual variability, and support rapid response. Consequently, embedding design values into both product development and standard-setting fosters trust, usability, and inclusivity. When standards are treated as adaptive, human-centered frameworks, they enable experimentation, innovation, and continuous improvement.

Institutional and cultural support is critical. Policymakers, designers, regulators, and manufacturers must collaborate, fund design research, pilot flexible compliance models, and share data openly. By embracing this approach, we move from mere compliance to a holistic understanding of safety: one that is inclusive, responsive, and capable of adapting to new challenges. Safety becomes more than technical protection; it becomes experiential, relational, and empowering, safeguarding not only the body but also the human experience.

This is a call to action: protective systems should foster trust, dignity, and meaningful change, reflecting the evolving realities of high-risk environments. By centering human experience alongside regulatory rigor, design-led approaches demonstrate how adaptive standards, informed by lived experience, can generate safer, more inclusive, and forward-looking solutions for the future of high-risk sport.

10.4 Recommendations for Future Research and Practice

Future work should focus on developing and testing prototypes that incorporate advanced materials, wearable sensors, modular padding, and quick-release mechanisms. These innovations have the potential to enhance both performance and safety, allowing protective equipment to respond intelligently to impact and environmental conditions.

Expanding stakeholder involvement will also be critical. Beyond riders and designers, collaboration should include trainers, stewards, manufacturers, and regulatory authorities to ensure that new solutions reflect the realities of racing environments and operational demands. Such broad engagement encourages a shared sense of ownership and accountability across the industry, strengthening both design outcomes and adoption.

Longitudinal studies are needed to evaluate redesigned vests over multiple racing seasons, capturing data on durability, comfort, and real-world safety performance. Insights from this research could also inform the adaptation of similar technologies in other high-impact sports, creating opportunities for interdisciplinary innovation that extends beyond equestrian contexts.

Finally, partnerships with regulatory bodies will be essential to revise existing standards, ensuring they remain flexible and responsive to technological change. In parallel, ongoing research should investigate how digital technologies (including real-time monitoring systems, automated emergency alerts, and post-event data analysis) can be integrated to transform safety management into a proactive, data-driven process. Together, these initiatives outline a pathway toward protective systems that are not only compliant, but genuinely intelligent, adaptive, and human-centered.

Table 3 highlights the key differences between current safety vest designs and the proposed D-UX improvements. Using human-centered design principles, the new model improves comfort, adaptability, and protection, overcoming limitations of existing models and supporting a more inclusive approach to rider safety.

Table 3. Comparison of Current vs. Proposed vests features

Feature / Attribute	**Current Vest Design (Limitations)**	**Proposed D-UX Improvements**
Comfort	Poor fit, especially for female jockeys; rigid padding limits movement.	Ergonomic, gender-inclusive design with flexible, adaptive padding for better fit and comfort.
Modularity	Fixed padding configuration; no adaptability to user needs or activity type.	Interchangeable, modular padding tailored for different impact zones and riding styles.

Smart Materials & Technology	Limited breathability; no use of responsive or digital technologies.	Integration of impact-absorbing foams and smart textiles (e.g., Kevlar®, Coolmax®); optional sensor systems for monitoring performance and impact data.
Emergency Accessibility	Rigid closures make medical access difficult during emergencies.	Quick-release mechanisms designed for rapid and safe removal by medical personnel.
Feature / Attribute	Current Vest Design (Limitations)	Proposed D-UX Improvements
Comfort	Poor fit, especially for female jockeys; rigid padding limits movement.	Ergonomic, gender-inclusive design with flexible, adaptive padding for better fit and comfort.

This table outlines the main differences between existing safety vest designs and the proposed D-UX improvements. By applying HCD principles, the proposed model enhances comfort, adaptability, and protection, addressing limitations in current models and promoting a more inclusive approach to rider safety. Besides, it demonstrates that by incorporating ergonomic, modular, and technology-driven features, protective equipment can move beyond compliance-focused designs toward systems that empower users and respond effectively to real-world safety needs.

10.5 Final Reflections

Designing for safety is inseparable from designing for human experience. Protective equipment must be usable, inclusive, adaptive, and responsive. By combining human-centered design, participatory approaches, and flexible standards, we can create systems that not only safeguard bodies but also empower users, enhance performance, and support well-being in complex, high-risk environments. Innovation in safety extends beyond the development of new materials, sensors, or technical regulations. It is grounded in collaboration, empathy, observation, and continuous learning. The principles explored throughout this book provide a roadmap for designing safer, smarter, and more inclusive equipment. These principles offer models that can be applied across sports, workplaces, and daily life, showing how thoughtful design can transform ordinary protective measures into systems that respect and respond to human needs.

The journey traced in this book highlights that safety is not merely a technical requirement but a human-centered practice. Users are not passive recipients of equipment or standards. They are active co-creators who shape technologies through their lived experiences, insights, and collaboration. Standards, when treated as living frameworks, become tools that guide innovation rather than barriers to progress. Design, informed by empathy, adaptability, and inclusivity, transforms protective equipment into systems that protect, empower, and honor the dignity of those who rely on it.

This approach emphasizes that safety and human experience are inseparable. When equipment is designed with attention to comfort, usability, inclusivity, and responsiveness, it strengthens trust, encourages adoption, and supports resilience. Co-dependent users, such as medical staff, teammates, or colleagues, become integral to safety outcomes, highlighting the importance of considering broader networks of interaction. Recognizing diverse bodies, roles, and operational contexts ensures that protective systems are not only effective in principle but effective in practice, responding to real-world challenges and diverse needs.

Institutional and cultural support is essential for these principles to take root. Designers, regulators, policymakers, manufacturers, and end-users must collaborate to fund research, pilot innovative approaches, share data openly, and maintain adaptive standards. Through this collaborative commitment, safety moves from compliance-oriented measures to holistic systems that foster trust, care, and empowerment. Safety becomes relational, experiential, and forward-looking, supporting not just the protection of the body but also the flourishing of human potential.

Ultimately, this book demonstrates that the future of safety depends on the integration of human-centered design, adaptive standards, and participatory approaches. Protective systems can be more than technical requirements if they are developed with empathy, inclusivity, and responsiveness. They can anticipate change, respect dignity, and evolve alongside technological advances, societal expectations, and the diversity of users. By embracing this vision, we have the opportunity to create environments where safety, innovation, and human experience converge, ensuring that people are not only protected but also supported, empowered, and valued in every context of high-risk activity or daily life.

Safety is not only a condition to be achieved. It is a practice to be lived, shaped, and continuously improved through the interactions of people, technology, and standards. By centering human experience, fostering collaboration, and allowing flexibility within systems, we can realize a future in which safety is meaningful, inclusive, and capable of responding to the challenges and opportunities of tomorrow.

References

Acar, O. A., Tarakci, M., and Van Knippenberg, D. (2018). Creativity and Innovation Under Constraints: A Cross-Disciplinary Integrative Review. *Journal of Management*, *45*(1), 96-121. https://doi.org/10.1177/0149206318805832 (Original work published 2019)

Allen, R. and Sriram, R. D. (2000). The Role of Standards in Innovation. *Technological Forecasting and Social Change*, Vol. 64. Available from: https://tsapps.nist.gov/publication/get_pdf.cfm?pub_id=821473

Al-Bayati, A.J., Rener, A.T., Listello, M.P., Mohamed, M. (2023). PPE non-compliance among construction workers: An assessment of contributing factors utilizing fuzzy theory. *J Safety Res*. 2023 Jun;85:242-253. doi: 10.1016/j.jsr.2023.02.008. Epub 2023 Feb 18. PMID: 37330874.

Anderson, M. (2010). Review of the book Standards and Their Stories: How Quantifying, Classifying and Formalizing Practices Shape Everyday Life. *Journal of Interdisciplinary History 41*(1), 124-125. https://muse.jhu.edu/article/381684.

Aryana, B., Osvalder, A.-L., and Borell, J. (2024). Design strategies to reduce personal protective equipment noncompliance. I P.-L. Rau (Red.), *Ergonomics In Design* (Vol. 129, s. 57-66) https://doi.org/10.54941/ahfe1004811

Ashford, N.A., and Hall, R.P. (2018). Technology, Globalization, and Sustainable Development: Transforming the Industrial State (1st ed.). Routledge. https://doi.org/10.4324/9780429468056

ASTM International. (2023). ASTM F1937-23: Standard specification for body protectors used in horse sports and horseback riding. Available at https://store.astm.org/f1937-04r17.html

Babik, K. R., Alper, J., and Downey, A. S. (2023). *Personal protective equipment and personal protective technology product standardization for a resilient public health supply chain: proceedings of a workshop*. National Academies Press, Washington, DC. Available from: https://nap.nationalacademies.org/catalog/27094/personal-protective-equipment-and-personal-protective-technology-product-standardization-for-a-resilient-public-health-supply-chain

Bannon, L. (2011). Reimagining HCI: Toward a more human-centered perspective. *Interactions*, *18*(4), 50–57. https://doi.org/10.1145/1978822.1978 833

Berni, A., and Borgianni, Y. (2021). From the definition of user experience to a framework to classify its applications in design. *Proceedings of the Design Society*, 1, 1627–1636. doi:10.1017/pds.2021.424

BETA (2024). Body protection made easy. Available at https://beta-uk.org/body-protectors/

Bishop PA, Balilonis G, Davis JK, and Zhang Y (2013). Ergonomics and Comfort in Protective and Sport Clothing: A Brief Review. *Journal of Ergonomics*, S2:005. DOI: 10.4172/2165-7556.S2-005

Bjögvinsson, E., Ehn, P., Hillgren, P. -A. (2012) Design things and design thinking: Contemporary participatory design challenges. *Design Issues*, 28(3), 101–116.

Blind, K. (2016). Chapter 15: The impact of regulation on innovation. *In Handbook of Innovation Policy Impact.* Cheltenham, UK: Edward Elgar Publishing. Retrieved Sep 22, 2025, from https://doi.org/10.4337/9781784711856.00022

Braun, V., & Clarke, V. (2014). What can "thematic analysis" offer health and wellbeing researchers? *International Journal of Qualitative Studies on Health and Well-Being*, *9*(1). https://doi.org/10.3402/qhw.v9.26152

Brisbine, B.R., Radcliffe, C.R., Jones, M.L.H., Stirling, L., Coltman, C.E. (2022) Does the fit of personal protective equipment affect functional performance? A systematic review across occupational domains. *PLoS ONE* 17(11):e0278174. https://doi.org/10.1371/journal.pone.0278174

[BSI] British Standards Institution. (2025). *BS 30417:2025 Provision of Inclusive PPE – Guide*. Retrieved from https://www.bsigroup.com/en-GB/standards/bs-30417-2025-provision-of-inclusive-ppe-guide/

Brown, T. (2009). *Change by Design: How Design Thinking Creates New Alternatives for Business and Society*. Harvard Business Press.

Brown, T. and Katz, B. (2011), Change by Design. Journal of Product Innovation Management, 28: 381-383. https://doi.org/10.1111/j.1540-5885.2011.00806.x

Busch, L. (2011). *Standards: Recipes for reality*. Mit Press. https://doi.org/10.7551/mitpress/8962.003.0012

Chaffin, D.B., Andersson, G.B., and Martin, B.J. (2006). *Occupational biomechanics (4th Ed.)*. John wiley & sons.

Chen, D. -S., Lu-Lin, C., Hummels, C., Koskinen, I. (2016). Social design: An introduction. *International Journal of Design*, 10(1), 1–5.

De Vries, H. J. (1999). *Standardization: A business approach to the role of national standardization organizations*. Kluwer Academic Publishers, Boston, Dordrecht, London (1999).

Dempsey, P. G. (2023). Human Factors/Ergonomics Contributions to an Equitable Personal Protective Equipment Protections Strategy. *Proceedings of the Human Factors and Ergonomics Society Annual Meeting*, 67(1), 816-820. https://doi.org/10.1177/21695067231192598 (Original work published 2023)

Desmet, P., Porcelijn, R., and Van Dijk, M. (2007). Emotional Design; Application of a research-based design approach. *Knowledge, Technology & Policy, 20*(3), 141-155. DOI 10.1007/s12130-007-9018-4

DiMaggio, P. J., and Powell, W. W. (1983). The Iron Cage Revisited: Institutional Isomorphism and Collective Rationality in Organizational Fields. *American Sociological Review*, *48*(2), 147–160. https://doi.org/10.2307/2095101

Dobbin, A. (2024). Jockey vest review remains hidden amid renewed push. Racenet – Latest News. 11 June 2024. Retrieved from https://www.racenet.com.au /news/the-move-by-hong-kong-to-green-light-use-of-japanese-safety-vest-leads-to-renewed-push-in-australia-20240610

Doorn, N., and Hansson, S.O. (2015). Design for the Value of Safety. *In: Van den Hoven, J., Vermaas, P., van de Poel, I. (eds) Handbook of Ethics, Values, and Technological Design*. Springer, Dordrecht. https://doi.org/10.1007/978-94-007-6970-0_19

Dorries, B. (2025). 'Ridiculous': Melbourne Cup-winning jockeys demand vest action. Racenet -Latest News 10 July 2025. Retrieved from https://www.racenet.com. au/news/ridiculous-trio-of-melbourne-cup-winning-jockeys-demand-action-on-safety-vest-review-20250710

Dorst, K. (2011) The core of 'design thinking' and its application. *Design Studies*, 32(6), 521–532. doi: 10.1016/j.destud.2011.07.006

Dunne, L. (2010). Smart Clothing in Practice: Key Design Barriers to Commercialization. *Fashion Practice*, *2*(1), 41–65. https://doi.org/10.2752/175693810X12640026716393

Duryan, M., Smyth, H., Roberts, A., Rowlinson, S., Sherratt, F. (2020). Knowledge transfer for occupational health and safety: Cultivating health and safety learning culture in construction firms. *Accid Anal Prev*. 2020 May;139:105496. doi: 10.1016/j.aap.2020.105496. Epub 2020 Mar 18. PMID: 32199157.

Fargnoli, M. (2021). Design for Safety and Human Factors in Industrial Engineering: A review towards a unified framework. *In Proceedings of the 11th Annual International Conference on Industrial Engineering and Operations Management*, Singapore (Vol. 7). March 2021. Available from https://www.ieomsociety.org/singapore2021/papers/1307.pdf

Fleck, J., and Howells, J. (2001). Technology, the Technology Complex and the Paradox of Technological Determinism. *Technology Analysis & Strategic Management*, *13*(4), 523–531. https://doi.org/10.1080/09537320127285

Follett, H. (2025). PPE Should Fit Everyone: Inclusivity in Protective Clothing. Respirex Blog. Retrieved from https://blog.respirex.com/blog/ppe-should-fit-everyone-inclusivity-in-protective-clothing?

Foote, C.E., McIntosh, A., V'Landys, P., Bulloch, K. (2011). Health and Safety in Australian horse racing. Rural Industries Research and Development Corporation. RIRDC Publication No 10/067, RIRDC Project No. PRJ-000765. Retrieved from https://www.agrifutures.com.au/wp-content/uploads/publications/10-067.pdf

Foote, C.E., Gibson, T.J., McGauran, P.J. (2014). (2014). Evaluation of Safety Vests: Health and Safety in Australian Racing. Health and Safety in Australian Racing. RIRDC Publication No. 14/037, 22 April 2014. Retrieved from https://agrifutures.com.au/wp-content/uploads/publications/14-037.pdf

Forlizzi, J. (2007). The product ecology: Understanding social product use and supporting design culture. *International Journal of Design, 2*(1), 11-20

Forlizzi, J., and Battarbee, K. (2018). Understanding Experience in Interactive Systems. In Proceedings of the 5th conference on Designing interactive systems: processes, practices, methods, and techniques, DIS2004, August 1–4, 2004, Cambridge, Massachusetts, USA. https://doi.org/10.1184/R1/6470606.v1

Foster, Nancy (2025). Can Collaborative Efforts to Improve Device Design Improve Safety? *American Hospital Association Blog*. 28 March 2025. Available from https://www.aha.org/news/blog/2025-03-28-can-collaborative-efforts-improve-device-design-improve-safety

Garrett, J. J. (2011). *The Elements of User Experience: User-Centered Design for the Web and Beyond* (2nd ed.). New Riders, Berkeley, CA.

Giacomin, J. (2014). What Is Human Centred Design? *The Design Journal*, *17*(4), 606–623. https://doi.org/10.2752/175630614X14056185480186

Giusti Gestri, Lisa (2019). Safety vests for jockeys: a case study of primary and dependent-secondary users affecting the evolution of vest design in the Australian horse-racing industry. Swinburne. Thesis. https://doi.org/10.25916/sut.26296477.v1

Giusti Gestri, L., and Barnes, C. (2020) Product standards as a barrier to innovation: the case of jockey's safety vests. In Boess, S., Cheung, M. and Cain, R. (eds.), *Synergy - DRS International Conference 2020*, 11-14 August, Held online. https://doi.org/10.21606/drs.2020.251

Giusti Gestri, L. (2023a). Is It Still Worth Challenging the Status Quo? A Design Process Article With an Example Application to Australian Jockey's Safety Vests. *Ergonomics in Design*, 31(2). https://doi.org/10.1177/1064804620986642

Giusti Gestri L (2023b) Wearable technology may assist in reducing jockeys' injuries if integrated into their safety vests: a qualitative study. *Front. Sports Act. Living* 5:1167110. doi: 10.3389/fspor.2023.1167110

Giusti Gestri, L. (2024). Assessment of Design Values in Designing Product Standards: A Design Study on a Sport PPE. *International Journal of Sports Science and Physical Education*, 9(1), 15-25. https://doi.org/10.11648/j.ijsspe.20240901.13

Göhlich, D., Bender, B., Fay, T.-A., and Gericke, K. (2021). Product requirements specification process in product development. In the *Proceedings of the Design Society*, 1, 2459–2470. doi:10.1017/pds.2021.507

Gould, S. MD; Roy, I. R. MPH; Gabriel, F. Dib MS; Lowe, R. NRP; Lin, CY. MD, FACSM (2025). Equestrian Sports Safety: An Update on Safety Equipment and

Event Medical Coverage. *Current Sports Medicine Reports* 24(1):p 11-17, January 2025. | DOI: 10.1249/JSR.0000000000001222

Gumusel, B. (2025). Why design needs theory: The Key to Meaningful and Effective Design Solutions. *Learn architecture – Articles*. 14 April 2025. Retrieved from https://learnarchitecture.net/articles/7099-why-design-needs-theory.html

Gunningham, N. (2007). Designing OSH standards: process, safety case and best practice. *Policy and Practice in Health and Safety*, *5*(2), 3–24. https://doi.org/10.1080/14774003.2007.11667691

Hall, A., Ferrarello, L., Anderson, P., Cooper, R., & Ross, C. (2019, September). Designing Design for Safety: How emergent methods indicate new safer future design practices. I*n the International Association of Societies of Design Research Conference 2019*. Available from https://researchonline.rca.ac.uk/4076/3/IASDR%20Designing%20Design%20for%20Safety%20V3%20FINAL%20%281%29.pdf

Hartley, J., Sørensen, E., Torfing, J. (2013) Collaborative innovation: a viable alternative to market competition and organizational entrepreneurship. *Public Administration Review*, 73(6), 821–830. doi: 10.1111/puar.12136

Hassenzahl, M., and Tractinsky, N. (2006). User experience - a research agenda. *Behaviour & Information Technology*, *25*(2), 91–97. https://doi.org/10.1080/01449290500330331

Henderson, R. M., and Clark, K. B. (1990). Architectural Innovation: The Reconfiguration of Existing Product Technologies and the Failure of Established Firms. *Administrative Science Quarterly*, *35*(1), 9–30. https://doi.org/10.2307/2393549

Hockey, G. R. J., Gaillard, A. W. K., and Burov, O. (Eds.) (2003). *Operator functional state: the assessment and prediction of human performance: degradation in complex tasks*. IOS Press.

Hokka, J. (2023). Gender and the Diversity of the Human Body as Challenges for the Inclusive Design of Wearable Technology. *Fashion Practice*, *16*(1), 108–133. https://doi.org/10.1080/17569370.2023.2250153

Horberry, T., and Burgess-Limerick, R. (2015). Applying a Human-Centred Process to Re-Design Equipment and Work Environments. *Safety*, 1(1), 7-15. https://doi.org/10.3390/safety1010007

IEEE (2018). Why Technical Standards are Essential in Product Development. IEEE Innovation at Work. 29th March 2018. Available from https://innovationatwork .ieee.org/why-technical-standards-are-essential-in-product-development/

[ISO] International Organization for Standardization (2019). ISO 9241-210:2019. *Ergonomics of human-system interaction. Part 210: Human-centred design for interactive systems.* International Organization for Standardization. Available from: https://www.iso.org/standard/77520.html

Janson, D. (2023). *Gender and Sex Bias in PPE Provision.* Doctoral thesis. University of Bath. Available from https://researchportal.bath.ac.uk/en/studentTheses/gen der-and-sex-bias-in-ppe-provision

Jiancaro, T. (2018). Empathy-Based Design Approaches. *In: Filimowicz, M., Tzankova, V. (eds) New Directions in Third Wave Human-Computer Interaction: Volume 2 - Methodologies.* Human–Computer Interaction Series. Springer, Cham. https://doi.org/10.1007/978-3-319-73374-6_4

Kampfmann, T., Bernert, P., Lang, D.J. (2023).Toward a modular evaluation approach of real-world laboratories: Findings from a literature review, *Research Evaluation*, Volume 32, Issue 1, January 2023, Pages 128–143, https://doi.org/10.1093/reseval/rvac029

Kim, Y.J. (2025). A Comprehensive Approach to Intelligent Garment for Posture Correction: A Preliminary Study for Integrating Aesthetics and Sensor Interaction. *J. Vib. Eng. Technol.* 13, 60 (2025). https://doi.org/10.1007/s42417-024-01738-w

ouprie, M., and Visser, F. S. (2009). A framework for empathy in design: stepping into and out of the user's life. *Journal of Engineering Design*, 20(5), 437–448. https://doi.org/10.1080/09544820902875033

Kundu, O., Uyarra, E., Ortega-Argiles, R., Tirado, M.M., Kitsos, T., Yuan, P.Y. (2024). Impacts of policy-driven public procurement: a methodological review, *Science and Public Policy*, Volume 52, Issue 1, February 2025, Pages 50–64, https://doi.org/10.1093/scipol/scae058

Lampland, M., & Star, S. L. (Eds.). (2009). *Standards and their stories: How quantifying, classifying, and formalizing practices shape everyday life*. Cornell University Press.

Law, EL.-C., Roto, V., Vermeeren, APOS., Kort, J., and Hassenzahl, M. (2009). Understanding, scoping and defining user experience: a survey approach. In S.n. (Ed.), *Proceedings of CHI 2009, the 27th Annual CHI Conference on Human Factors in Computing Systems* (pp. 719-728). ACM, New York.

Legg, K.A., Cochrane, D.J., Gee, E.K. *et al.* (2023). Physical Fitness of Thoroughbred Horse Racing Jockeys. *Journal of sci. In sport and exercise* (2023). https://doi.org/10.1007/s42978-023-00257-6

Leonard, D., and Rayport, J.F. (1997). Spark innovation through empathic design. *Harv Bus Rev*. 1997 Nov-Dec;75(6):102-13. PMID: 10174792.

Lewis, J. M., Ricard, L. M., Klijn, E. H., Figueras, T. Y. (2017) *Innovation in City Governments: Structures, Networks, and Leadership* (New York: Routledge).

Lewis, J. M., McGann, M., & Blomkamp, E. (2020). When design meets power: design thinking, public sector innovation and the politics of policymaking. *Policy & Politics*, *48*(1), 111-130. https://doi.org/10.1332/030557319X15579230420081

Loibner, M., Hagauer, S., Schwantzer, G., Berghold, A., and Zatloukal, K. (2019). Limiting factors for wearing personal protective equipment (PPE) in a health care environment evaluated in a randomised study. *Plos One*, 14(1), e0210775. https://doi.org/10.1371/journal.pone.0210775

Malta, G., Matera, S., Plescia, F., Calascibetta, A., Argo, A., & Cannizzaro, E. (2024). Occupational accidents and the use of PPE: A global meta-analysis. *Frontiers in Public Health*, *12*, 1368991. https://doi.org/10.3389/fpubh.2024.1368991

Manders-Huits, N. (2011). What Values in Design? The Challenge of Incorporating Moral Values into Design. *Sci Eng Ethics*. 2011, 17, 271–287. https://doi.org/10.1007/s11948-010-9198-2

McCarthy, J., and Wright, P. (2004). *Technology as experience. Interactions*, Vol. 11, 5 (September + October 2004), 42–43. https://doi.org/10.1145/1015530.1015549

Molina, J.B., Sern, L., Ibrahimi, A., and Jarquin, C. (2024). *Designing for Diversity: The Importance of Inclusive Design in Sports*. WDO – World Design Organization. Available from https://wdo.org/designing-for-diversity-the-importance-of-inclusive-design-in-sports/

Morgan, H. (2024). Using Triangulation and Crystallization to Make Qualitative Studies Trustworthy and Rigorous. *The Qualitative Report*, *29*(7), 1844-1856. https://doi.org/10.46743/2160-3715/2024.6071

Nagare, A. (2024). Ergonomics and Usability in Wearable Tech: Balancing Aesthetics, Functionality, and Comfort through Industrial Design. *International Journal of Innovative Research In Technology (IJIRT)*, 11(6), 1482-1488.

National Academies of Sciences, Engineering, and Medicine. (2023). *Personal Protective Equipment and Personal Protective Technology Product Standardization for a Resilient Public Health Supply Chain: Proceedings of a Workshop*. Washington, DC: The National Academies Press. https://doi.org/10.17226/27094

Norman, D. A. (2004). *Emotional design: Why we love (or hate) everyday things.* Basic Books/Hachette Book Group.

Norman, D.A. (2013). *The design of everyday things (Revised and Expanded Edition).* MIT Press.

O'Connor, S., Warrington, G., McGoldrick, A., Cullen, S. (2017) Epidemiology of injury due to race-day jockey falls in professional flat and jump horse racing in Ireland, 2011–2015. *J Athl Train.* (2017) 52(12):1140–6. doi: 10.4085/1062-6050-52.12.17

Osborne, S., Brown, L. (2013) Introduction: innovation in public services. *Handbook of Innovation in Public Services* [Online], eds S. P. Osborne, L. Brown (Cheltenham: Edward Elgar Publishing), pp 1–11, www.elgaronline.com/view/9781849809740,xml.

Peng, Y., Dong, J., Gu, Y., Zhang, Y., Long, J., Park, S., Liu, T., and Huang, Y. (2024). Smart temperature-adaptive thermal regulation textiles integrating passive radiative cooling and reversible heat storage. *Nano Energy*, *131*, Article 110311. https://doi.org/10.1016/j.nanoen.2024.110311

Perego, P., Sironi, R., (2022). Full-stack User-Centered Approach for Wearable technology design. *International Journal of Design Sciences and Technology*. 2022, Volume 25 Number 1. ISSN 1630-7267.

Perera, C. (2006). Standardization in Product Development and Design. In *Hesser, W., H. de Fries, A. Feilzer Standardisation in Companies and Markets*. Hamburg: Helmut-Schmidt-University, pp. 141 – 176.

Pfau, T., Spence, A., Starke, S., Ferrari, M., and Wilson, A. (2009). Modern riding style improves horse racing times. *Science* (New York, N.Y.), 325(5938), 289. https://doi.org/10.1126/science.1174605

Racing Australia (2025). Australian rules of racing. *Racing Australia*, as at 1st May 2025. Retrieved from https://www.racingaustralia.horse/uploadimg/Australian_rules_of_Racing/Australian_Rules_of_Racing_01_May_2025.pdf

Riganelli Engineering (2025). The Critical Role of Standards in Product Development. *Riganelli Engineering - Product Development*. Available from https://www.riganelli.ca/articles/the-critical-role-of-standards-in-product-development

Rotinen, L. (2022). *Sustainable public procurement of personal protective equipment in healthcare*. Lappeenranta–Lahti University of Technology LUT Master's Programme in Strategy, Innovation and Sustainability, Master's thesis. https://urn.fi/URN:NBN:fi-fe2022090156971

Sanders, E. B. N., and Stappers, P. J. (2014). Probes, toolkits and prototypes: three approaches to making in codesigning. *CoDesign*, *10*(1), 5–14. https://doi.org/10.1080/15710882.2014.888183

Sarkki, S., Heikkinen, H.I., Komu, T., Partanen, M., Vanhanen, K., Lépy, É. (2020). How boundary objects help to perform roles of science arbiter, honest broker, and issue advocate, *Science and Public Policy*, Volume 47, Issue 2, April 2020, Pages 161–171, https://doi.org/10.1093/scipol/scz055

Schilli, B., Kern, I. (2006). Importance of Product Design and Data Exchange Standards in the Product Life Cycle. *In: Prozessorientiertes Product Lifecycle Management*. Springer, Berlin, Heidelberg. https://doi.org/10.1007/3-540-28442-7_14

Schimpf, S., Weber, H., Gerlach, T. (2024). Enabling radical and potentially disruptive innovations through interdisciplinarity: challenges and practices in industrial companies. *J Innov Entrep* **13**, 45 (2024). https://doi.org/10.1186/s13731-024-00402-7

Schonheyder, J.F. (2019). *Method Development for the Design of Safety-Critical Systems: The Space between Design Research and Professional Design Practice*. Oslo School of Architecture and Design. Doctoral Thesis. Available from: http://hdl.handle.net/11250/2602820

Shi, J., Li, H., Xu, F., Tao, X. (2021). Materials in advanced design of personal protective equipment: a review. *Mater Today Adv.* 2021 Dec;12:100171. doi: 10.1016/j.mtadv.2021.100171. Epub 2021 Sep 8. PMID: 34514364; PMCID: PMC8423993.

Smith, B. (2018). Generalizability in Qualitative Research: Misunderstandings, Opportunities and Recommendations for the Sport and Exercise Sciences. Qualitative Research in Sport, Exercise & Health, 10, 137-149. https://doi.org/10.1080/2159676X.2017.1393221

Soltani, A. M., and Pouypouy, H. (2019). Standardization and regulations of nanotechnology and recent government policies across the world on nanomaterials. In *Advances in phytonanotechnology* (pp. 419-446). Academic Press. https://doi.org/10.1016/B978-0-12-815322-2.00020-1

Stilgoe, J., Owen, R., & Macnaghten, P. (2013). Developing a framework for responsible innovation. *Research Policy*, Elsevier, vol. 42(9), pages 1568-1580. DOI: 10.1016/j.respol.2013.05.008

Tamjid,M.I., Abtew, M,A., Kopot, C. (2025). Lightweight Textile and Fiber-Reinforced Composites for Soft Body Armor (SBA): Advances in Panel Design, Materials, and Testing Standards. *Journal of Composites Science*. 2025; 9(7):337. https://doi.org/10.3390/jcs9070337

Tang, K.H.D. (2025). Advances in Thermoregulating Textiles: Materials, Mechanisms, and Applications. *Textiles*. 2025; 5(2):22. https://doi.org/10.3390/textiles5020022

The Hong Kong Jockey Club (2005). Equipment register (equine and human). V 5.1, February 2025. Retrieved from at: https://racing.hkjc.com/racing/english/racing-info/rules_pdf/Equipment-Register-Full-202502-English.pdf

Tractinsky, N, Katz, A.S., Ikar, D. (2000). What is beautiful is usable, *Interacting with Computers*, Volume 13, Issue 2, December 2000, Pages 127–145, https://doi.org/10.1016/S0953-5438(00)00031-X

Van der Velden, M., and Mörtberg, C. (2015). Participatory Design and Design for Values. *In: Van den Hoven, J., Vermaas, P., van de Poel, I. (eds) Handbook of Ethics, Values, and Technological Design*. Springer, Dordrecht. https://doi.org/10.1007/978-94-007-6970-0_33

Verganti R. (2009). *Design - Driven Innovation. Changing the rules of competition by radically innovating what things mean*. Harvard Business Press.

Von Hippel, E. (2005). *Democratizing innovation*. The MIT Press.

Webster, J. (2010). *The perception of comfort and fit of personal protective equipment in sport*. Loughborough University. Thesis. https://hdl.handle.net/2134/6242

Yin, R. K. (2018). *Case Study Research and Applications: Design and Methods* (6th ed.). Thousand Oaks, CA: Sage

Zhang J, Johnson TR, Patel VL, Paige DL, Kubose T. (2003). Using usability heuristics to evaluate patient safety of medical devices. *J Biomed Inform*. 2003 Feb-Apr;36(1-2):23-30. doi: 10.1016/s1532-0464(03)00060-1. PMID: 14552844

www.ingramcontent.com/pod-product-compliance
Lightning Source LLC
LaVergne TN
LVHW022107200926
848481LV00003B/1089